Stories from the MAINE COAST

SKIPPERS, SHIPS *and* STORMS

HARRY GRATWICK

FOREWORD *by* PHILIP CONKLING

Published by The History Press
Charleston, SC 29403
www.historypress.net

Cover image: Collision between *Little Annie* and *Virginia*, 1903. Anonymous.

First published 2012

Manufactured in the United States

ISBN 978.1.60949.249.6

Library of Congress Cataloging-in-Publication Data

Gratwick, Harry.
Stories from the Maine coast : skippers, ships and storms / Harry Gratwick.
p. cm.
Includes bibliographical references.
ISBN 978-1-60949-249-6
1. Atlantic Coast (Me.)--History. 2. Seafaring life--Maine--History. 3. Ships--Maine--History. 4. Shipbuilding--Maine. 5. Ship captains--Maine--Biography. I. Title.
F27.A75G727 2012
974.1--dc23
2011047612

Notice: The information in this book is true and complete to the best of our knowledge. It is offered without guarantee on the part of the author or The History Press. The author and The History Press disclaim all liability in connection with the use of this book.

*This book is dedicated to Francis "Mac" McAdoo,
friend and skipper extraordinaire.*

CONTENTS

Foreword

Geologists call the Maine coastline a "drowned coast," not because so many of its mariners have drowned here (as my young son once supposed), but because after the last glacier melted, the sea flooded what was once a rolling coastal lowland to create a deeply incised coastline and a long chain of isolated islands. This geological history also bequeathed us a bountiful legacy of coves, harbors, guts and guzzles from which mariners have launched uncountable thousands of vessels, underscoring Maine's maritime heritage since earliest times.

Maine's intricate island and coastal topography, and the deep-water channels connecting them, offered—almost begged—its settlers to become seafarers. For three hundred years, the first (and only) highways that connected the Maine coast and archipelago to the routes of international commerce were maritime sailing routes that originated in isolated places like Carver's Harbor on Vinalhaven and the Deer Isle Thorofare that streamed by the salty, spruce-lined coves of Stonington.

Island settlement was also spurred by the legendary concentrations of spawning cod and haddock that congregated around the shores of Maine's islands in such vast numbers that places like the Isles of Shoals and Haddock Ledge in Penobscot Bay were named for the bounty created by their annual migrations and made fishermen out of settlers who originally came here to farm.

The mariners' stories that Harry Gratwick has assembled in this book include accounts of some of the great triumphs, as well as the tragedies, that maritime life leaves in the bones of those "who follow the sea." Since I happened to have married into a family that occupied a magnificent four-story house built in 1857 by the seafarer and successful fish merchant Timothy Lane and his wife, Rebecca, for their nine children on a hilltop overlooking the busy commercial Carver's Harbor on Vinalhaven Island, I have a particular admiration for this collection of great Maine sea stories. Few of our family members, nor our innumerable visitors during the past forty years who have spent a night in Timothy and Rebecca Lane's solidly built ten-bedroom house, have failed to sense in the creaks of this wonderful structure the spirit of one of Maine's great marine innovators who, with generations of other nautical families along Maine's long and lonely coast, has left us with such a rich maritime legacy. Thankfully, this volume nicely captures Maine's maritime spirit.

Philip Conkling
Founder and President
Island Institute in Rockland, Maine

Acknowledgements

A number of people have been invaluable in helping me with information and advice on writing this book. I am indebted to Mac and Cynthia McAdoo for their PT boat memories, as well as their friendship; to Emily Lane and Dick and Lee Morehouse for their stories about the Lane family and Lane's Island; and to Peter and Lucy Bell Sellers, from whom I learned about boat building persistence. And thank you to John Washburn for his friendship and memories of Hurricane Edna.

Special thanks to Joey Zehr and Frank Garber for their technological wizardry, to Mark Warner for his materials on the *Royal Tar* and to George Buzby for being a friend and an old salt.

The Rhodes 19 story would not have been possible without assistance from Dave Whittier, Peter Plumb, Stuart Scharaga, Jim Taylor, Fred Brehob, David Pyles and my brother, Joel Gratwick.

I am grateful to Cipperly Good and Donald Garrold at the Penobscot Marine Museum, to Nathan Lipfert at the Maine Maritime Museum, to Erik Fleischer at the John Fitch Steamboat Museum, to Bill Haviland at the Deer Isle–Stonington Historical Society, to Eric Jergensen and Janice Zenter at the Maine Maritime Academy, to Philip Conkling and Gillian Thompson at the Island Institute and to Richard Lindermann at the Bowdoin College Library Special Collections.

Kudos to Sue Radley, Bill Chilles, Roy Heisler and Loretta Chilles at the Vinalhaven Historical Society, as well as Valerie Morton and Linda Whittington at Vinalhaven Public Library, for their generous help and support.

Finally, the encouragement and enthusiasm of my son, Stephen, and the love and editorial skill of my wife, Tita, are deeply appreciated.

I

SKIPPERS

THE LANE FAMILY: A TRADITION OF SEAFARERS

The roots of the Lane family reach well back into history. According to James Fitts,* author of the *Lane Genealogies*, a member of the Lane family fought with Oliver Cromwell in the English Civil War, and a Lane came over on the *Mayflower*. Later Lanes fought in the French and Indian War, as well as in the Revolutionary War battles of Lexington and Concord and Bunker Hill.

In less conflicted times, Mr. Fitts informs us that Lanes distinguished themselves as physicians, lawyers, inventors, politicians, teachers, ministers, missionaries, businessmen and writers. However, by far the greatest number of the North American Lane family have made their living from the sea; many were fishermen, shipbuilders, shipowners and ship's masters. In his *Genealogies*, Fitts lists fourteen sea captains and another five who were shipowners.

Some Prominent Ancestors

LEVI LANE (1754–1806) was born near Gloucester, Massachusetts, and began his career as an apprentice sailmaker. At the Battle of Bunker Hill,

---

* Unless otherwise noted, passages in quotations are from Fitts.

he saw action as a member of Captain Nathaniel Warner's company. Levi then went to France and returned on the same ship as the Marquis de Lafayette in 1777. After the war, Mr. Lane became "a sail maker of wide repute. His [sails] were the best to come out of Boston." Lane went on to become a leading merchant and shipowner, although, Fitts writes, "he lost some vessels to French spoliation [plundering] about 1800." He was a pew owner and prominent member of the First Universalist Church in Boston.

CAPTAIN FRANCIS LANE (1756–1829) was a minuteman (colonial militia) from Gloucester who fought at the Battle of Bunker Hill. For the remainder of the war, Lane served on board a privateer, for which he received a share of prize money. After the war, he studied navigation and was subsequently appointed master of a ship that made numerous voyages to the West Indies. When his ship was wrecked, he managed to recover the cargo (of cotton). He then returned to New England, where he "coasted [traded] between Boston and Portland for many years."

GIDEON LANE (1764–1821) was also a mariner from Gloucester. During the War of 1812, a British force entered the town's harbor with the intention of destroying American shipping. Lane's seventeen-year-old daughter, Clara, appealed to the British commander to spare her father's schooner, *Federalist.* Fitts tells us, "The lieutenant in command looked at Miss Lane for a few moments then declared he could not resist the request of such a pretty lass and would leave the vessel unmolested." Lane was known for his diligence and energy as a businessman, as well as for his cheerful disposition. He died in Gloucester at the age of fifty-seven after a long illness, which, his obituary noted, "he endured with remarkable fortitude and patience."

GUSTAVUS ADOLPHUS LANE (1811–1878) also continued the tradition of Lane seafarers. Fitts refers to him as "a master mariner from an old Gloucester family." Lane joined a ship's crew and soon rose to the rank of captain, making many "foreign voyages." For a while, however, Captain Lane took time off from his maritime career and "was employed in the granite business." His obituary notes, "A serious illness compelled his retirement from active service in the prime of life."

JACOB LANE (1800–1882) was an exception to the tradition of Lane seafarers. Jacob settled in Minot, Maine (near Lewiston). He was described as a "tall, muscular man, standing six feet three in his stocking feet and straight as an arrow." Up to "four score years of age," he challenged his sons and the young men of their generation to "lay him on his back." At one point in his life, Jacob Lane was a captain of a military company noted, not surprisingly, for its "good discipline and fine appearance on muster days." Jacob Lane was frequently urged to take town office but always declined. For most of his life, he was "a successful schoolteacher."

This brings us to the Lanes of Vinalhaven. ISACHAR (b. 1760) and his younger brother, BENJAMIN (b. 1762), both grew up in Gloucester, thoroughly steeped in the family's seafaring tradition. As young men, they moved to Penobscot Bay's Fox Islands shortly after the Revolution. It was a propitious time, since in 1789 the town of Vinalhaven was incorporated.

In those early years of Vinalhaven's history, men often sought their wives on Matinicus; Isachar and Benjamin were no exception. Benjamin married Margaret Hall, who bore him seven children. (The Hall family apparently had an ample supply of marriageable daughters.) We know little about

Lane family plot, Lane's Island, Vinalhaven. Captain Timothy Lane's white stone is seen on the left. *Author's collection.*

Isachar Lane except that he married Margaret's sister, Susan, and that he lost a hand in a hunting accident. Eventually, Isachar left Vinalhaven and "followed the sea."

Benjamin Lane settled his family on Griffin's (Lane's) Island, which he purchased from another early settler, Thaddeus Carver, for a new milch cow and a heifer. He lived there until his death in 1842. He and Margaret had two daughters and five sons, two of whom—Joseph and Timothy—would be successful mariners/businessmen.

"The Appearance of the Island Is Not Very Inviting"

With these rather indecisive words, the *Rockland Democrat & Free Press* described Vinalhaven for potential settlers in 1859. The article added:

> *Its outer edge is a succession of granite hills and coves, the coves making up between the hills and furnishing fine and numerous harbors for fishing craft. Where the land is not cleared a low growth of spruce has taken root. These places contain the residences of the inhabitants and are cultivated as well as are the sea-going habits of their owners.*

The newspaper article noted that although there was some farming, fishing was the most important industry in the town. By the mid-nineteenth century, there were between seventy-five and one hundred vessels engaged in fishing, employing 600 men, 280 of whom were listed as "master mariners." Vinalhaven's fishermen took distant voyages to the Grand Banks and the Gulf of St. Lawrence for cod and mackerel, an operation that kept four freighters busy carrying cargoes of fish from the island to Boston.

In the early nineteenth century, the next generation of Carvers and Lanes took the lead in Vinalhaven's rapidly growing fishing and shipbuilding industries. One of the leaders was Thaddeus Carver's son Reuben, who, with his brother John, built the 146-ton *Plymouth Rock* in 1826, the first of twelve schooners they would ultimately construct. Although Reuben was primarily a shipbuilder, his other interests included lumbering, operating a sawmill, constructing buildings and curing and smoking fish.

A recent Vinalhaven Historical Society newsletter noted, "The Carver brothers were not the only major sea captains on the island; the Lane family wealth also came from the sea." Benjamin Lane's youngest son, Timothy, was a good example of a successful Vinalhaven mariner/businessman. The youngest Lane was born on the island in 1805 and shipped out as a common seaman at the age of twenty-one. He soon became a mate, and within a few years, young Lane was master of a one-hundred-ton schooner that he had built and partially owned.

In 1833, Timothy Lane's life changed dramatically. He quit his life as a sea captain, married Rebecca Smith from Vinalhaven and built a house on Lane's Island. The couple would have four children. Lane entered the fishing business as part owner of a "coasting" schooner. From these modest beginnings, he rose to be principal owner of "fifteen to twenty" trading and fishing vessels. The largest, the 119-ton schooner *Rebecca C. Lane*, was built in Boston in 1864 and named for his wife.

At the same time, Captain Timothy, as he was known, together with his older brother Joseph, established a thriving fish-curing operation on the shores of Lane's Island. The Lane brothers also added to their income by opening a store on the island that "furnished outfits for fishing vessels."

Through his various business enterprises, Timothy Lane rose to become one of the wealthiest citizens on Vinalhaven, which by mid-century numbered over two thousand people. In 1859, for example, it was estimated that Lane's shipping interests alone grossed approximately $20,000.00. By 1865, the value of Timothy Lane's property and business ventures were such that he paid a tax of $1,328.73, "the largest amount ever assessed against any one person up to that time," as reported by James Fitts in *Lane Genealogies.*

Lane's Hall

With so much of his wealth coming from the sea, it is surprising that Timothy Lane listed his occupation simply as "farmer" in an early island census. Lane's Island is approximately seventy-five acres, from which Timothy had thirty-five to fifty tons of hay cut annually. In the middle of the nineteenth century, he built what was described as "a princely residence on a beautiful spot. The southern side looks out to the broad Atlantic and the front faces the packet [ship] as she makes for Carver's Harbor."

Over the years, four additional houses were built on the island for members of the Lane family. Captain Timothy was as hospitable to his

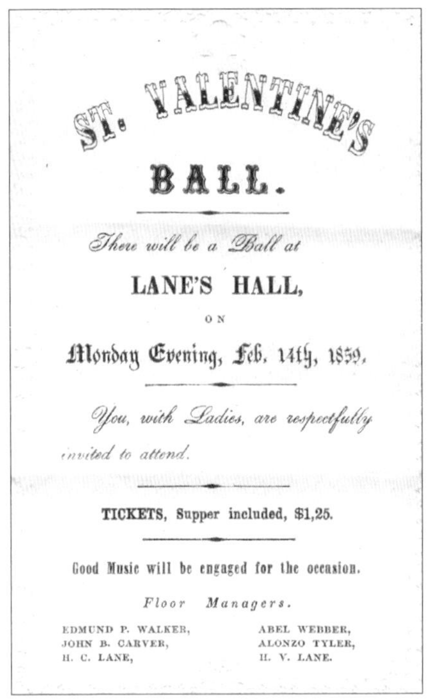

ST. VALENTINE'S

BALL.

There will be a Ball at

LANE'S HALL,

ON

Monday Evening, Feb. 14th, 1859.

You, with Ladies, are respectfully invited to attend.

TICKETS, Supper included, $1,25.

Good Music will be engaged for the occasion.

Floor Managers.

EDMUND P. WALKER,
JOHN B. CARVER,
H. C. LANE,
ABEL WEBBER,
ALONZO TYLER,
H. V. LANE.

A St. Valentine's Day party invitation to Lane's Hall, 1859. *Courtesy of the Morehouse family.*

business contacts as he was generous to his family. Reportedly, he was so welcoming to visiting captains and customers that his wife, Rebecca, never knew how many would be coming for dinner.

The house, known at the time as Lane's Hall, was the scene of many parties and galas. The St. Valentine's Ball on February 14, 1859, was a particularly memorable event. Tickets were $1.25 and included supper and dancing on the vast third floor, which ran the length of the large house.

Captain Timothy Lane died in 1871, although his enterprises continued to prosper. For years, his son Francis returned each season with his twenty-one-ton schooner, *Willow*, filled with codfish and mackerel. And in 1878, the Lane & Libby Fish Co. opened for business. By the 1880s, it had grown into one of the largest fishing operations in New England, with many modern innovations such as icehouses and a hydraulic press for packing fish.

Lane's Hall became Ocean House in 1874 when Timothy's widow, age seventy-three, and her son Francis M. Lane opened the building as a commercial rooming establishment. Rebecca died in 1888 at the age of eighty-one, although Francis kept the hotel open into the 1890s. Apparently, the retreat attracted actors—Otis Skinner among them—who occasionally performed in village productions.

An excerpt from a letter written in 1880 by Rebecca L. Littlefield gives us a description of Ocean House:

> *As visitors sail into Carver's Harbor they will observe opposite the Boat Wharf a large boarding house for the accommodation of tourists or pleasure seekers who, during the heat of the summer, seek the comfort conducted by sea air and the inspiration of ocean scenery.*

Lane's Hall from a nineteenth-century drawing. *Courtesy of Vinalhaven Historical Society.*

> *This large and commodious house is located upon a small island recognized as Lanes Island, so called from the name of its former owner Capt. Timothy Lane, who died nine years ago, who was an active business man and prominently noted for his kindness of heart, and who, with his equally kind-hearted wife, in addition to their own four children, adopted and raised five other children.*

The Rockaway Inn and Beyond

From the mid-1890s to 1928, the old house languished with only a few occupants until it reopened as the Rockaway Inn. The new owners were Bruce Grindle and his mother. Bruce described the inn in a letter to a prospective visitor:

> *Rockaway Inn is located on Lanes Island and connected with Vinalhaven by a bridge. It is far enough removed from town to insure quiet and perfect restfulness.*

Rockaway Inn, 1920s. *Courtesy of Vinalhaven Historical Society.*

> *The panoramic view from the Inn is unexcelled by any along the Atlantic Seaboard. A cool breeze is enjoyed from this vantage point during the warmest days of summer. Bathing and boating are pastimes entered into by our guests.*
>
> *All rooms are outside rooms and on the same floor with the baths. The rooms are spacious and furnished in such a manner as to please the most exacting. A large open fireplace, lending to the New England home atmosphere, sets off the Lounge room, a masterpiece in the art of woodworking.*
>
> *The Inn opens May 1st and closes October 1st. Our rates are $25 per person per week, or $4 a day.*

Astrid Rosen Philbrook worked at the Rockaway Inn from 1939 to 1942. Astrid told me she wanted a job so much that she lied about her age, since she was only thirteen when she started. When she confessed later to Una Jones, one of the proprietors, Una said, "I knew that, but you wanted the job so bad I just said yes."

Astrid said she was paid six dollars a week and began work early each morning. She cleaned the living room and dining room, set the table, made the beds (there were seven bedrooms), waited on table and did the dishes. Astrid told me that since she was the only helper, she asked if she could bring a friend to help. Her friend was hired and got paid three dollars a week for doing the dishes.

"The same clientele tended to come every year. They came from all over," Astrid said, "even during the war [World War II]." The guests were mostly older people who enjoyed the beach, took walks and relaxed. There was one family with two boys who came for a month. At the end of the time, they gave her a tip of forty dollars, which she loved. Astrid told me that Fred and Una Jones ran the place until it closed in 1956.

In the 1960s, Dick and Lee Morehouse were looking for a place they could buy on Vinalhaven and where their children and grandchildren could visit them in the summer. At one point, a member of the Nature Conservancy mentioned the long-vacant Rockaway Inn. He told Dick Morehouse, "You probably aren't interested in this old wreck, but I'll show it to you anyway."

As they walked around the old inn, what amazed the Morehouses was that for a house that had been abandoned for so long, it hadn't been trashed. Dick told me, "About the only damage we found was a squirt of toothpaste running along one of the halls. Can you imagine the condition a vacant house would be in today?" he remarked wistfully at one point.

Dick and Lee Morehouse on Lane's Island, circa 1975. *Courtesy of the Morehouse family.*

I recently asked Dick, who is an architect, to describe the condition of the house when they first saw it. Dick said the windows were covered with plywood that the Nature Conservancy had nailed on for protection. The net effect, however, was that the interior of the house couldn't breathe, and all the wallpaper had fallen off the walls.

Dick also told me he was amazed to find the old furniture from the Rockaway Inn still in good shape. "It was from the Paine Furniture Co. in Boston and was, and still is, very heavy. I guess it was too heavy for intruders to move out of the house," he mused. Dick said the old inn was heated. "We found a giant steam furnace in the basement and radiators in all ten of the bedrooms."

The Morehouses bought the house from the Nature Conservancy in 1967 and, with the help of their family, have brought it back to life. The upkeep is endless, but they have had lots of help. One of the first projects was to wallpaper the bedrooms, which Lee Morehouse and her sister did, two or three rooms at a time, every summer.

In 1967–68, a large part of Lane's Island was threatened with commercial development. A campaign was launched to raise funds to preserve the island's remaining fifty acres as a town park. Vinalhaven residents, who for years had enjoyed the rocky headlands, rolling moors and island's wildlife, signed a petition urging the town to purchase that portion of the island threatened by development. A committee was formed, and the necessary funds were raised, with the assistance of the Nature Conservancy.

Today, the Lane's Island Preserve covers two-thirds of the island. Many people come to picnic, paint, bird-watch or enjoy the beach. Others walk the numerous trails that lace the preserve, many of which lead to the shore and dramatic views of the sea. With several thousand visitors annually, it is safe to assume that the hospitable and generous Timothy Lane would be delighted with the way his island is being used today.

THE COASTAL TRADER: HENRY LUFKIN OF DEER ISLE

Henry Lufkin was born on Deer Isle in 1797, the son of parents who migrated from Gloucester, Massachusetts. In 1844, Lufkin began a diary, which he kept until 1863. Lufkin's diary is valuable as the log of

Captain Henry Lufkin transported cargo up and down the East Coast. *Courtesy of Deer Isle–Stonington Historical Society.*

a coastal trader and also provides us with a look at mid-nineteenth-century life on a Maine island.

As one might expect, weather and health concerns were paramount to Captain Lufkin. We also learn that his life consisted of a good deal more than skippering one of his small fleet of two-masted coastal schooners. The Lufkin family (Henry had two brothers) were shipbuilders as well as sailors. In addition, Henry was a prosperous farmer, a skilled blacksmith, a carpenter and a shoemaker.

For most of the year, Captain Henry, as he was called, sailed his various ships from as far Down East as Petit Manan and Machias to as far south as New York and occasionally Virginia. A partial list of Maine ports where he traded includes Bangor, Mt. Desert, Sedgwick, Castine, Fox Islands (Vinalhaven), Thomaston and Portland.

As Clayton H. Gross points out in his book *Island Chronicles*:

> *Since no road system worthy of the name existed at the time, coastal shipping supplied not only personal transportation, but also made it possible for a number of local industries to flourish. Coasters, though not fast, provided cheap, low cost transportation, which made Deer Isle products competitive.*

Lufkin's ships carried a variety of goods on his coastal runs. His diary makes references to cargoes of lumber, granite, cotton, rum, lobsters and coal, although the following entries indicate that he seems to have mostly carried timber and coal:

> *Bargained for a load of timber for a wharf…Engaged a freight of wood at $3.50 a cord…Dropped off the brig in the harbor and went to Scotts Island and cutting timber…Finished loading 143 tons of coal…got underway for Roundout (a coal port in Virginia) with coal at $1.43 per ton…Took in 2,000 bricks, hoop poles and staves.*

Main Street, town of Sunset on Deer Isle. Lufkin's house is in the distance on the left. *Courtesy of Deer Isle–Stonington Historical Society.*

There was no such thing as a typical voyage for Captain Lufkin. His notes while on the schooner *Banner*, in the spring of 1847, give us an idea of what his schedule was like, although we are not told of his specific cargo:

> *April 17th, Got underway for Boston (from Deer Isle), breeze increased and returned.*
> *19th, Got underway at 8 a.m. come to Camden 2 p.m.*
> *20th, Beat down to Owls Head, calm, come to.*
> *21st, Calm, underway at 3 a.m. and drifted in a light wind to White Head (an island off Tenants Harbor). Fog, thick, rain.*
> *22nd, Thick fog; got underway, squally.*

This went on until *Banner* arrived at Portland on April 23. Lufkin left Portland that afternoon in a "nasty storm" and six days later arrived in Boston, where he made the cryptic notation, "Tired enough."

Although he was obviously exhausted, Captain Lufkin left the next day for Nantucket, where another of his ships, *Susan and Jane*, lay damaged. Lufkin writes that he "employed all hands in ships duty and got the water

Henry Lufkin's house in Sunset. *Courtesy of Deer Isle–Stonington Historical Society.*

out." *Susan and Jane* was subsequently hauled out for repair, which took several days.

Lufkin was not above occasionally adding personal observations: "William [a crew member] wants to write to his intended, but is rather bashful. Pains me all over, not very smart." And he was upset by the proliferation of sand flies in Nantucket, "enough to put out your eyes."

Back at Deer Isle on June 3, Captain Lufkin headed for Machias, arriving on the eleventh. After "running ashore on the flats" on June 20 while entering Jonesport, he took the rest of the month to get to New York, where he arrived on June 30. Captain Henry spent the remainder of what appears to have been a typical summer hauling cargo up and down the East Coast.

During the summer of 1847, he made two other notations that are of interest. In July, while in New York, he reveals, "Had a watch stole last night." And in August, Lufkin got word that his ship *Banner* "was lost on Matinicus." "Didn't care for her, let her go," was his reaction. (He had never liked the ship because it was a poor sailer.)

The winter of 1847–48 was an unlucky one for Lufkin personally. A December 21, 1847 entry reads: "At three this morning a schooner run across my bow and carried away martingale [a stay or a spike bowsprit] and jib boom. His anchor across ours and we couldn't get underway until the tide slackened in the evening." And on January 1, 1848, "George Canigan deserted. My chest has been broken into and between $75–$100 taken out. Search has been made to find him, but have not succeeded."

Death and Dying

During the winter months the death rate appeared to increase in coastal communities, and Lufkin's diary is filled with references to this. November and December 1846 were particularly distressing times. On November 30, he writes, "Bad news that Francis Haskell and crew are lost on little Cranberry Island." The next day, Lufkin notes, "Sad news that the brig *Lincoln* is lost and four of her crew drowned." (The *Lincoln* was a Deer Isle ship that foundered off Martha's Vineyard.) And on January 20, 1847, Lufkin writes, "News of Capt. Thomson's death." Apparently, Captain Dudley Thompson had been murdered in a mutiny at sea, though given the irregular delivery of mail, the news did not reach Deer Isle for several months.

On January 2, 1847, he writes, "Mr. Down very sick." (He died the next day.) On February 14, he notes, "Mr. Sylvester very sick and not very well myself." (Mr. Sylvester also died the next day.) Probably the worst disaster Lufkin records that winter occurred when Captain Rufus York's ship, en route for Rockland, caught fire and the elderly York and one of his sons perished. Lufkin's terse entry reads, "Sad accident happened to Capt. York, burnt."

Henry Lufkin was a temperate, God-fearing man. Death was a part of life, and he believed in the immortality of the soul. Nevertheless, the illness and death of his wife of thirty-four years, Mehitable Saunders Lufkin, in March 1853 is revealed as a heart-wrenching experience. Beginning in January, his diary is filled with references to her declining health:

> *Wife very sick, grows bad fast this evening, small hopes of her recovery… Wife no better, swell bad…Wife grows worse, she is in great distress, swelled bad legs and body…Mother is no better in pain and distress and sick to her stomach…Doctor has been here five times…Mother is not better but grown worse, bad cough and swells more…The rest of us is all tired and have bad colds.*

And on March 1, 1853, he writes:

> *This is a terrible day to us all. Mother is terrible sick, until she took her flight to the unknown world at eight in the evening. It is a solemn time with us, may God sanctify it to us all for everlasting good, may it serve to lead us to the rest that is laid up for the people of God.*

After gathering his children, Lufkin planned to bury his wife on March 5, but twelve inches of snow fell, also keeping the minister from Sedgwick away. The next day, March 6, Lufkin records, "A beautiful day for the funeral, warm as summer, a great many people here this day. We have laid wife and mother in the grave, never to see her again. It is a solemn time. May God seal it upon our hearts that it will do us good."

Last Years

Captain Henry's younger brother Joseph died in 1851, and by 1853, he makes a series of references to the increasing mental disorder of Joseph's wife, Sally Jo. "Sally Jo is crazy as a coot...Sally Jo is bad...Made a straightjacket for Sally Jo...Sally Jo is hollering all the time...Sally Jo is gone to the hospital."

What is interesting about this is that families at the time normally kept mentally unstable relatives out of sight—in a closet, a cellar or an attic—in an effort to avoid public disgrace. Lufkin had tried to keep Sally Jo at home, but when he realized her situation was hopeless, she was put in a hospital. In this case, it was the Insane Hospital at Augusta, which had opened in 1840. Remarkably, Sally Jo lived until 1884.

In February 22, 1854, Henry Lufkin's diary enigmatically and pragmatically records, "Married this day." He does not tell us whom he married, but Lufkin family history reveals the bride was Sarah Dodge, who was sixteen years younger than the fifty-six-year-old captain. (I should add that for the previous year, Lufkin had a series of housekeepers, including a daughter, none of whom worked out.)

Henry Lufkin lived until 1868, continuing his career as a coastal trader. When on Deer Isle, his diary informs us that he helped build a new schoolhouse to replace the one that had burned down. He writes that "John Small has hauled his house," which is a reminder that house moving was common until the early twentieth century. In Lufkin's

House hauling in the nineteenth century was often cheaper than building a new one. *Courtesy of Vinalhaven Historical Society.*

time, when labor was cheap and materials expensive, it was often more economical to move a house rather than build a new one. His diary also reports that by the mid-nineteenth century, steamships were in evidence. The steamer *Rockland* ran a regular route from Rockland to Machias until it was destroyed during the Civil War.

There were signs, however, that the old sailor was beginning to feel his age: "I am not well, Johnson hauling wood." Nevertheless, the last entries in his diary make references to "Captain Henry…cutting timber and hauling wood."

The Sailing Master: Edward Nichols of Searsport

Edward Nichols was one of a renowned group of nineteenth-century Searsport sailing masters. In his book *Searsport Sea Captains*, Frederick Black writes, "Searsport furnished more masters to the square mile than any other community in the United States. Searsport masters were in great demand and they commanded vessels built in every port on the Atlantic coast."

Black gives us a list of the prominent Searsport sailing families, who include fifteen Pendletons (his mother's family) and twenty-seven Nicholses, so we know Edward came from good seafaring stock.

Captain William Nichols was one of a distinguished group of nineteenth-century Searsport sailing masters. *Courtesy of Penobscot Marine Museum.*

Nichols was the son of William Nichols and Nancy Pendleton. He was born in 1844, and in 1873, he married Martha Mills from Bangor. Nichols had three daughters, one of whom was born at sea in 1876, while he was captain of the bark *Clara*. Edward Nichols was a devoted husband and father, and his wife and two older daughters accompanied him on many of his voyages.

During the course of his extensive career as a sailing master, Nichols became famous as the publisher of the *Ocean Chronicle*, the only newspaper known to have been printed on a sailing vessel. Nichols began his publishing and sailing master's career on the bark *Clara* in 1878, but in 1880, *Clara* was wrecked off the coast of South Africa. Nichols then transferred to the ship *Frank Pendleton* and, at the urging of his wife, changed the name of his publication from *Pill Garlic* to *Ocean Chronicle.*

Ocean Chronicle was at the same time a newspaper, a captain's diary and a ship's log. Nichols comes across as a man with a sense of humor, a sense of duty and a purveyor of folk wisdom. He is part Garrison Keillor, part Will Rogers and part Benjamin Franklin. Nothing is sacred in *Ocean Chronicle.* Nichols takes on all comers. In particular, he was enraged by the obsolete and archaic restrictions on American shipping, which will be discussed in due course.

On his travels, Nichols described the places he visited, the people he met and his musings while at sea. In *Ocean Chronicle*, we are introduced to his favorite ports. At the top of the list was Rio de Janeiro: "The harbor is one of the finest in the world, the natural scenery surrounding it is unsurpassed in grandeur." San Diego's harbor was also "one of the finest," and he raved about the climate. He was also impressed with Hong Kong, "although their port charges are rather high."

Nichols visited Australia several times, and he was particularly impressed with Melbourne: "They have many things to be proud of. They boast of their fine harbor, cleanliness of their city, their broad streets, their modern conveniences and their general hospitality. I think Melbourne is one of the finest cities I have ever visited."

For fifteen years, Nichols sailed his ships around the world and in the process crossed the Equator sixty-one times. One might ask what cargo *Frank Pendleton* was hauling all those miles. In fact, the two hundred-foot *Frank*, as the ship was affectionately called, could carry almost two thousand tons of cargo, which was most often coal.

In 1888, Nichols wrote, "Having had three cargoes of coal on board in twelve months, coal dust has found its way into every nook and corner of the *Frank*." All was not misery, however. In 1891, Nichols added, "I do not feel very *grand*, in bringing *sand*, from Valparaiso to New York, but I am told that the *Frank* is the first that ever did it."

Government Regulations

As he circled the globe, Captain Nichols became increasingly irate over arcane American shipping regulations. He was also concerned with the plight of American seamen. Beginning in 1887, his opinions were reflected in a series of articles in *Ocean Chronicle*.

That spring, Nichols wrote two columns bemoaning the fact that American shipping "was on its last legs." In the article, Nichols pointed out that United States ships carried less than 15 percent of the world's trade, and the percentage was declining every year due to what he considered defective legislation passed by Congress that was "mutilated with amendments."

Nichols cited examples: when an American ship "hits the water, all the birds gather together for a taste of the carcass; the broker, ship chandler, carpenter, sail-maker, butcher and blacksmith. Then down swoops the American Eagle [U.S. government] to fill her rapacious maw." Nichols asked the question, "What does the Government do for ships? This can be answered in four words. 'Nothing for; lots against.'"

In a later *Ocean Chronicle* piece, Nichols focused on the declining numbers of American seamen. "The American seamen, where are they? They are among the things of the past. The United States at present could not furnish enough seamen to man a fleet." The angry captain went on to deplore the

CHRONICLE.

PUBLISHED BY E. P. NICHOLS.

AT SEA, ON BOARD SHIP, "FRANK PENDLETON."

This is printed for PASTIME ONLY, and sent to friends as a letter, therefore not open to criticism. *TERMS; ONE LETTER*

VOL. 39. JULY. 25, 1883. NUMBER 8.

THE OCEAN CHRONICLE.

This number will be printed the same as the former numbers, — for amusement. Although the printing of this sheet cause some remarks to be made, such as "It is boyish," "What is the good of it?" "It is a queer 'pastime'," &c. yet, I shall keep on until I tire of it. Nearly every Capt. has something to do to pass away his spare time at sea, and that which would be pleasant for one, would be distasteful to another. Some make ships models and rig them; some work with the scroll-saw; some make fancy mats: some, spend their time in reading, but each pleases himself, as to what he shall do. I enjoy working with type, and cannot see that it is any more 'foolish' than others pastime, although the contents may be void of sense.— it is my *letter* to friends.

E. P. Nichols.

THE "FRANK PENDLETON."

To my friends who have never seen the ship will give a brief description of her in her present situation—June 14, 1883. lat. 43 N, lon 4 W. The ship is situated in the Atlantic Ocean. She is bounded on the east, by a north east wind, on the south, by a head-sea; on the west, by coal-dust; and on the north, by fish-line and log. Her length over all is two hundred feet, from the root of her nose to the tip of her tail, and her nose is fifty four ft, six in. long. She is twenty five feet through the pit of the stomach, and thirty nine across the chest. Her legs are twenty two feet long, and her arms, —of which she has several,—vary in length, from forty to eighty feet. On full stomach, she carries about 1950 tons.

There is quite a village on deck, a plan and directory of which can be seen at the Capt's office on the summit of Mount Lofty. When coming on board by the starboard gangway, you find yourself in Pendleton Square, from which are two streets leading south, called East and West Streets. In the middle of the southern side of the Square, is a Pavilion, out of which there is a passage leading to a dining hall and lodging house, which is also the head quarters of the municipal officers. In walking along East Street, you pass along the foot of Mt. Lofty and soon come to a private park, in the centre of which is a building for the protection of seamen, from the scorching rays of the sun, that they do not tan; from the midnight dew, the pouring rain, and from the chilling blast of the N. W. wind. From this park are steps leading up to Mt. Lofty, the top of which is flat, and the sides perpendicular. It has a railing around it, and on each side there is a place for an automatic elevator which all on board carry with them. The Capt's Office before mentioned as being on this Mount, is furnished in a very convenient manner. It has a good chart-table, several racks for charts, a compass, day and night-glasses, book-case with books, washstand and furnishings, a bed on which the Capt. can look out for the ship while he sleeps, and last is a box of tobacco and pipes.

After coming down from Mt. Lofty, by the way of the elevator on the northern side, and crossing Pendleton Square, you go down on to Shellback Common. From the Common north, are two streets, East, and West Jack St. Facing the common are a row of dwellings, in one of which the carpenter lives. In the next, the side-lamps, and other luminaries are kept, the next is inhabited by the cook. On Jack Street opening on to East, and West, you first come to the bakery and cook-shop, where they chew the hash with a knife all ready for swallowing. Next, is the carpenter shop where they make chips for kindling. Farther on you come to the habitation of the mariner, where in former times, interesting anecdotes of the "last ship" were told, but later years, an account of how much coal they shoveled, and how hot it was. The next place is the sail room, where the Frank's wardrobe is kept when she is not dressed. After passing this you come to a street running across from East to West Street, called Cross Street. On the north side of Cross, is the henery, under the management of the steward. Along each side of the henery is an alley, Paint Alley on one side, and Junk Alley on the other, each being thickly settled on one side, but the space between the two alleys forward of the henery, is the windlass space. All the houses have the name of the occupant over the door, except in case where one is well known, as is Walter Cousins, then the initials only are used.

This is only an outline description of the surface, but there is too much coal-dust down below to ask you to go down there, but if you would like to go up stairs, you are at liberty to do so, and you will there find many articles which will lead you to suppose that a surgeon stops there, as you will see trusses, slings, bands, leeches, yard*arms*, head and foot or sails, legs— of buntlines,— and eyes,— of the rigging, and many classes are represented up in the attic, for there is a good part of a ladies outfit, as you will find stays, braces, collars, caps, lacings, ties, pins and when it is blowing heavy there is generaly a great *bustle* aloft.

We will not try to give any more particulars about the ship, but will endeavor to find something written with better taste.

FRIENDSHIP.

It is friends, that make life worth living for. One who has no friends, though he may have all the money he may wish, can never be happy, while another who subsists on the bare necessities of life, having the good-will of his fellow-beings will lead a life of contentment. In most cases, "money makes friends" but what is such friendship worth? the name "Friendship"; is an *alias* for "Selfishness." A man with much money may have his friends, and retain them while his capital lasts, but let him lose his wealth, his power, and those who professed to be his friends will be the most exacting at the settlement. There are those who profess to be friends, and will try to impress upon your mind, their object is to aid and assist, but should you fail to make a return of three to them to one for yourself, they are your friends no longer. There are those who will invest their money with a hope that it will yield a profitable return, and if it fail to do so, and they are satisfied one has done his utmost to achieve a gain and failed, will try again; these are friends not to be forgotton; but when friends turn their capital from a channel that is sure to give them a good return, to another, where there is only a *hope* that it will yield them an equal gain; it is done out of friendship, and as such, cannot be too highly appreciated, and one with such friends,—though misfortune befall him— with health remaining, has good capital on which to start anew and should do his utmost to fulfill the "hope" of his friends, for in their hope lie his success, and their hope realized, his success is sure. It is not necessary one should have money, to be a friend, for in many ways one may act a friend by deeds of kindness that cost nothing and bind you closer to them than though they had used money for you and assumed a cool indifference to the comfort of you, or yours. Those who smile at your success and sympathize with your misfortune, are friends which you should highly esteem.

Front page of Nichols's newspaper, *Ocean Chronicle*. This was the only newspaper known to have been printed on a sailing vessel. *Courtesy of Penobscot Marine Museum.*

fact that "nine-tenths of the seamen sailing in American ships were foreigners without the intention of becoming a citizen."

In response to the clamor from shipping interests, and in an effort to aid American seamen, Congress passed the Dingley Shipping Bill in 1884. One aspect of the bill that Nichols liked reduced the exorbitant fees charged by boardinghouse keepers for mariners who were living ashore. He railed against other provisions in the bill that continued to discriminate against American seamen: "A sea-faring man is not a free citizen of the United States, for he is denied all rights and privileges of Civil Law, is not allowed to make a contract except as dictated by Government, and is only allowed to receive the wages due him at the will of Government."

Scientific Observations

Captain Nichols frequently added interesting scientific notes to the *Ocean Chronicle*. Several examples follow:

> *A queer phenomenon of the weather occurred as we were sailing along on an east by south course. With the wind northwest, all that portion of the sky on our starboard was dark and threatening and, at times, it rained in torrents. While on our port it was fine, clear weather. It lasted this way for two days.*
>
> *One who follows the sea often sees many things, which would be of great interest to men of science. On a dark night the surface of the water was illuminated by millions of lights which shown forth. A scoop net was taken and several fine specimens were soon on deck. They were semi-transparent and about the same consistency as a jellyfish. The following day we placed them in a dark room in a bottle and they were still luminous, but the day after they gave forth no light.*
>
> *Another phenomenon, which is witnessed at sea, it might be more proper to class it "atmospheric" than "oceanic"…St. Elmo's Fire shows itself on very dark nights, in thunderstorms. The appearance of this very light is as if a star was shining brightly, and the first thought is that the cloud has suddenly broken. While in the Pacific we had a most beautiful light, not only one, but many. The finest sight was on the mizzen.*
>
> *We will say a word about the "sea quake." There was but little wind, but a very heavy swell, which had been running for several days. When the first shock came it sounded as if several heavy barrels were let to roll over the deck. The second shock followed in about three seconds and this shook things up properly, as if the whole cargo were adrift. Another peculiar thing was the swell that had been running for days commenced to fall at once.*

Two Revolutions

In 1887, twelve years before the Spanish-American War, Edward Nichols wrote about one of his dreams:

> *Exciting news! Revolution in Cuba, the American flag hauled down and torn to shreds. The lives of all the foreign population are in danger. Each nationality has appealed to their respective governments for protection. The British and the Germans sent gunboats and Washington gave orders to send assistance to our citizens.*

Nichols's dream continued:

> *The gunboat* Sioux *could not be gotten ready for at least ten days and the* Delaware *would take twenty days to put in order. The Government after finding that there was not an American ship available had chartered four coal hulks to carry troops, provisions and ammunition and they will be towed to the scene of action in a few days. We awoke to find that all that was wanting to make our dream true, was—the revolution.*

Although this wasn't exactly what happened in 1898, Nichols's dream was nevertheless amazingly prescient.

Captain Nichols sailed his ship, *Frank Pendleton*, all over the world. The ship was in Valparaiso Harbor when a revolution broke out in Chile in 1891. *Courtesy of Penobscot Marine Museum.*

The other revolution Nichols writes about actually happened. When the ship *Frank Pendleton* entered Valparaiso Harbor in January 1891, Nichols wrote that he saw numerous ships at anchor, but the "pilot, broker, butcher and ship's chandler were conspicuous by their absence." Finally, a boat approached with Ed Griffin, a friend from Searsport, on board. Griffin told him there had been a revolution in Chile "and that they might or might not get probed by a cannon ball before morning."

Nichols and other American captains spent the next several months waiting for the situation to resolve itself. Hearing that there was bitter fighting in the north, one wag commented, "If we wait long enough they will have enough dead Chileans to make you a load of guano." Finally, in June, Nichols received instructions to leave if there was no business. He immediately had ballast (in this case sand) put on board and left for New York. The revolution ended a few months later with the victory of the Chilean navy over the Chilean army and with the president of Chile committing suicide.

Late in his career, Nichols wrote:

> *I have been many places and formed many pleasant acquaintances, all of which have been very enjoyable. The question has often been asked of me, "of all the places you have been which would you have chosen for a home"? My answer has always been; "In our own dear state of Maine, for there is no more beautiful scenery to be found than on the Penobscot Bay and River. Maine is good enough for me."*

THE PT* BOAT COMMANDER: FRANCIS MCADOO

Vinalhaven summer resident Francis McAdoo was born in New York City in 1916. Francis, known to his friends as "Mac," graduated from Princeton in 1938 and attended Columbia Business School for a year before joining the United States Navy. When I asked him why he picked the navy, he told me his father and two uncles had been in the navy in World War I. "My father told me, you'd better join the navy or you'll get drafted into the army."

Because his father was an avid sailor, Mac did a lot of small boat sailing during summers spent in Narragansett, Rhode Island. As a teenager, he also developed a taste for speedboats, and as a young man he owned a twenty-

* "PT" stands for patrol torpedo.

two-foot Chris-Craft, which he occasionally took across Narragansett Bay to Newport. It was kind of a mini PT boat and could go thirty to thirty-five knots. The result was that Mac learned a lot about small boat handling even before he joined the navy. (Mac's Maine boating experiences will be discussed in due course.)

With war looming and realizing that "Hitler was for real," Mac enlisted in the navy in 1940. He told me, "The navy was giving ensign's commissions to college grads in those days. You became an officer first and were trained afterward." Mac was sent to a three-month officers' training program at Northwestern University, where he learned naval regulations and skills like gunnery and navigation. "On-the-water training was impossible. There were eight hundred of us in each class."

The Background

When Japanese forces drove General Douglas MacArthur from the Philippines in May 1942, Lieutenant Commander John Bulkeley evacuated the general in an old seventy-foot-long British MTB (motor torpedo boat). It

PT boat on patrol somewhere in the South Pacific. *United States Navy photo.*

was a rough, three-day, seven-hundred-mile trip to southern Mindanao, and apparently MacArthur was seasick the entire voyage. From there, he was flown to Darwin in northern Australia.

MacArthur, however, was so pleased with the performance of the old MTB that he told the navy, "We need a lot of these boats to work in the islands." The British had introduced the prototype PT boat, which they used against German E boats in the English Channel. PT boats, as they came to be known, were good because they had a lot of armament and little draft.

Back in the United States, the navy decided to build a larger and faster boat. The Elco Boat Company, in Bayonne, New Jersey, designed an eighty-foot boat with three 1,500-horsepower Packard engines. Top speed for an Elco PT boat was fifty knots under normal sea conditions.

PT boats were very seaworthy because the three thousand gallons of fuel were located below the waterline in self-sealing tanks. Furthermore, fuel could be transferred from one tank to another. "They were wonderful, solid boats," Mac said. Armament included four torpedoes, a twenty-millimeter Swiss Oerlikon cannon and twin fifty-caliber machines guns.

Training

At the end of Mac's training at Northwestern, Lieutenant Commander John Bulkeley arrived and gave the class a pep talk about serving in PTs. Mac, who was about to be promoted to lieutenant junior grade, told Commander Bulkeley that he'd turn down his promotion if he could be assigned to a PT squadron. "Things were changing so fast after Pearl Harbor, I didn't know what I was getting into. I was still a lowly ensign and didn't know anything."

In September 1942, Mac got orders for further training in motor torpedo boats in Newport, Rhode Island. Again, there were so many students that they didn't get much on-the-water experience running the boat. "We were simply passengers," Mac said.

Lieutenant Francis McAdoo. *Courtesy of the McAdoo family.*

PT boat trainer near Newport, Rhode Island. *Courtesy of Elco Naval Division Electric Boat Company, Bayonne, New Jersey.*

The trainee class had a lot of fun, even though they didn't learn much about operating PT boats. Once, when they were heading down the bay, a torpedo was fired by mistake and hit a navy merchant ship, which promptly sank in the mud. "Boy, there was hell to pay," Mac recalled.

At the end of three months, Mac got orders to report to PT Boat Squadron Three at the Brooklyn Navy Yard. Mac was assigned to PT 129, with a crew of two officers and nine men. Boat and crew were loaded on board an oil tanker along with five other PT boats and their crews. The tanker sailed on December 24, 1942. "We didn't know where we were going, except that it was somewhere in the Pacific."

Before proceeding across the Pacific, PT Boat Squadron Three was dropped off at the Panama Canal for further military exercises. When it was seen that the engines couldn't go fast enough in hot weather, it was decided to add a third propeller to each boat.

A few weeks later, Squadron Three was reloaded on another tanker and taken, unescorted, across the Pacific. The hope was that, at a speed of fifteen knots and in a noncombat zone, the tanker would be able to avoid contact with the enemy.

After an uneventful eighteen-day crossing, they stopped briefly at Brisbane, Australia, before going on to Cairns. From there, they headed north to a PT boat base at Milne Bay, on the eastern tip of New Guinea.

The trip to New Guinea was beyond a PT boat's fuel range, so freighters in a Dutch convoy towed them. Mac remembers it was a sickening feeling to be towed. When the sea got rough, their bows went down, and the boats swung back and forth.

Mac told me at one point their towline broke, and when he came on deck, the convoy was nowhere in sight. They didn't have radar and were forbidden to use radio since they didn't want to alert the Japanese. Remembering his navigation training at Northwestern, Mac calculated their position and told his helmsman, "Take this course and go at thirty knots." Twenty minutes later, they spotted the convoy. This happened on Mac's twenty-seventh birthday.

On Patrol

By the middle of 1942, MacArthur's Third Army and the Australian army were pushing Japanese forces across the Owen Stanley Range, which ran the length of New Guinea. At that point, the Japanese high command attempted to reinforce its troops by sending a fleet of troop transports. MacArthur's response was a massive air attack, which sank the troop ships and drowned most of the reinforcements.

The remaining Japanese on New Guinea were becoming desperate for food. The area commander's solution was to bring supply ships to the north coast of the island. Supplies would then be offloaded into large steel barges, powered by an auxiliary engine and armed with a machine gun. The PT boat's primary mission was to keep the steel barges, also known as luggers, from landing. Normally they didn't waste torpedoes but tried to sink the barges with their machine guns and later with forty-millimeter cannons.

"We haunted the harbors where supplies were unloaded at night," Mac said. What made it all the more difficult, and dangerous, was that shore batteries also protected Japanese harbors. Operating at night often meant working in the moonlight, which PT boat crews hated because they could be seen.

PT boats were also vulnerable to Japanese fighter planes. Sometimes a plane would spot their wake and hit them from astern even though PT boat gunners were on watch. Enemy destroyers with radar were another problem. Mac said that one time a friend's boat was caught in a searchlight from a destroyer that had radar. They made a smoke screen to disguise their location and barely got away.

Mac told me that they usually went out every other night and in pairs. That way, if one boat got in trouble—i.e. hit a ledge—the other could pull it off. Failing that, they would destroy the PT and take off the crew. Part of the problem was that the navy was using old nineteenth-century

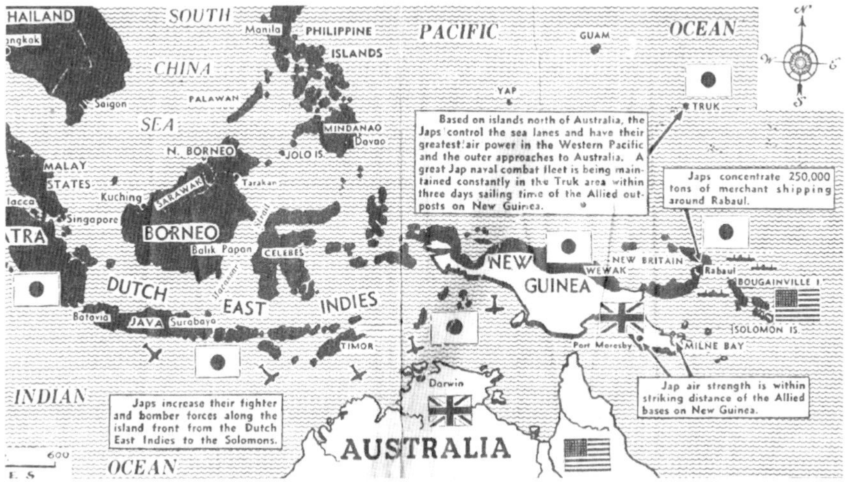

Areas of conflict in the South Pacific, from a World War II–era newspaper. *Courtesy of the McAdoo family.*

British maritime surveys for charts, which weren't very accurate. Nor did the century-old soundings show any reefs close to shore.

Usually, three PT patrols went out each night, a total of six boats. Their routine was generally two weeks on, one week off. Mac said there was always a lot of maintenance. Mechanics were constantly tinkering with the engines to make sure the boats operated at peak efficiency.

As the war progressed, the Japanese began to catch on to our tactics. Mac told me, "Once they got radar, we were much more vulnerable. Their destroyers could go forty-five knots and run us down, so we lightened our load by removing torpedoes and adding more deck guns." For their size, PT boats had more firepower than any other ship in the navy.

War Stories

Francis McAdoo's initial patrol occurred when PT 129 was sent out to destroy what was left of the Japanese transport fleet and pick up any survivors from MacArthur's air attack. "We were very excited heading out on our first patrol," he enthused.

Mac described what followed:

> *Thirty or forty miles off shore we found several Japanese swimming in the water. Some of my crew were gungho to start shooting them. In fact, a*

gunner's mate started to fire a Tommy gun until I stepped in. The men didn't like that, but they stopped, and began to get the reluctant Japanese on board.

Their prisoners were four "rather surly" young men, embarrassed that they had surrendered, even though they were way offshore with no chance of reaching the coast. Mac remembers being impressed with the tanned, muscled physical condition of his prisoners.

PT 129 returned to an advanced base, where the men could refuel and unload their prisoners. Three of the men got off, but the fourth stood in the bow and refused to move. He gestured to shoot him, thinking he was going to die anyway. At that point, a navy chief from the beach walked up to him and gently but firmly took his arm and walked him off the boat. "He just needed a nudge," Mac said.

Another time, Mac remembers a patrol with Annapolis grad Skip Dean, who was in PT 108. "We were both still a little green." It was a very dark night when their two boats crept quietly into a large harbor on the north coast of New Guinea. They kept their muffled engines at low speed to combat the strong current that kept sweeping them on shore. Dean finally

PT boat signaling while on patrol. *United States Navy photo.*

decided to anchor, and Mac took his boat out to patrol the harbor entrance. They stayed in touch by short-range radio.

Mac was several hundred yards away when the night was suddenly filled with tracer bullets. Was Dean being attacked? Mac rushed his boat back through the darkness, though he had to be careful what he shot at. It turned out that the Japanese had mistaken Dean's boat for a supply vessel and had sent two barges out to unload what they thought was a transport. As they approached, Dean cut his anchor line and opened fire on the barges at point-blank range.

Japanese shore batteries opened up, but both PTs got away in the confusion. The next day, Mac discovered that one of his engines wasn't functioning properly. When he had the boat inspected, mechanics found a pair of pants, presumably from a Japanese soldier, wrapped around the propeller.

Another time, Mac's PT 129 was the lead boat in an attack on enemy supply barges. The Japanese spotted them as they were coming up coast and got off a lucky shot. Mac's bow gunner and quartermaster, Harry Dhonau, was badly wounded in the leg by a twenty-millimeter shell. Although they tried to staunch the bleeding, Dhonau died before PT 129 could reach shore. Mac said, "I lost the best man on my boat."

As American marines drove the Japanese army back across New Guinea, PT boats had go considerable distances to get to the Japanese supply bases. Before one memorable patrol, they were told that no friendly ships were out there. "Anything moving would be the enemy."

Lieutenant Ed Farley, skipper of the lead boat, PT 220, had the latest radar. Around midnight, Farley radioed Mac to say that he had spotted three moving targets five miles ahead. Before attacking, Farley decided to call to confirm that they weren't friendly ships. There was no response to his first two calls.

Farley gave it one more try and radioed, "I'm going to fire if you don't identify yourself." Mac said they had cranked out their torpedoes and were ready to launch. Suddenly, after the third call, a voice blasted into their chart rooms: "Hold your fire, we are friendlies, en route to Buna." The blips on the radar screen were three old destroyer transports heading for a naval base at Dreggar Harbor. The radioman had just woken up and heard the final warning. "So much for naval intelligence," Mac said.

Another time, two PT boats had rescued a downed American flyer in a Japanese harbor that was mined and heavily guarded. One PT picked up the pilot while the other boat circled the harbor shooting up the Japanese shore defenses.

A few days later, Mac's friend Ed Hoagland was returning from a patrol and saw a squadron of American planes flying overhead. Ed radioed the other boat, "Don't shoot if they attack us but head for the beach. If we don't fire back they'll know we are friendly." Suddenly, the leading plane pealed off and dove toward them. At the last minute, it wagged its wings—a thank-you for having saved the downed flyer. Hoagland said it "scared the wits out of him" until he realized what was happening.

After eighteen months, Mac and the rest of Squadron Three were rotated back to the United States. Before he shipped out, Mac's commanding officer ordered him to take the mail to an island survey ship located about three hours north of New Guinea by boat.

They found the survey ship anchored in a lagoon surrounded by a coral reef with high mountains dominating the island. Mac told me he had never seen such a beautiful spot in his life. A picturesque native village was located near the site of an old prewar British coconut plantation. It reminded him of Bali Hai, the mythical Pacific isle. Mac said the whole scene "took my breath away." He determined to go back after the war, though he never made it.

Mac returned to the United States and served as executive officer of MTB Squadron Four in Melville, Rhode Island. Eighteen months later, he was back in the Pacific, where he commanded MTB Squadron Ten on an island off the western end of New Guinea. When he arrived, Mac found the squadron "in bad shape. The boats were badly in need of repair. By the time we were ready for reassignment," he told me, "the war was winding down."

Francis McAdoo mustered out as a lieutenant commander and was awarded the Silver Star Medal. The citation reads, in part:

> *For gallantry and intrepidity in action against the enemy. You took part in over thirty offensive patrols against Japanese barge traffic along the north coast of New Guinea. On 16 March 1943, your boat and MTB 114 engaged and sank six Japanese barges despite strong resistance by the enemy. Throughout, you displayed exemplary leadership and excellent seamanship in the execution of fearless and aggressive attacks against the enemy. Your actions are in keeping with the highest traditions of the United States Navy.*
>
> *Vice Admiral A.S. Carpendar*

Postwar

After the war, Mac returned to his day job as vice-president of the Emerson Drug Company, the family business in Baltimore, Maryland. In the 1950s, Mac resumed his friendship with John F. Kennedy, whom he had known before the war and whom he describes as "great fun." One time, when JFK was a senator, he and his brother Bobby came to Baltimore for a speed-reading course that Mac's company was hosting. After the session, Mac, JFK and Bobby sat around the boardroom. "Bobby, who was known to be difficult, complained, 'We aren't getting anything out of this and don't want to pay for it.'" Mac chuckled.

Mac never ran into Kennedy during World War II. JFK's PT squadron was stationed eight hundred miles east of New Guinea in the Solomon Islands. It is well known that the future president was almost killed in what one naval historian called "the most fouled-up PT operation in history." But that is another story.

For many summers after the war, Mac and his wife, Cynthia, chartered a variety of sloops and yawls for Mac's vacations. They cruised the area around Cape Cod for a few years before working their way up the coast and into Maine.

After years of chartering, Mac bought a twenty-seven-foot British-made sloop from an ex–PT man in Rockland, Maine. The McAdoos cruised in Maine waters for many years. As Mac said, "We first got to know the area from Kittery to Casco Bay." Later, they sailed to Mt. Desert, Cutler, Eastport and as far Down East as St. John. Mac told me that in those days they navigated mostly by dead reckoning, often operating in the fog.

Francis McAdoo cruising in Maine. *Courtesy of the McAdoo family.*

How did the McAdoos end up on Vinalhaven? Their connection was through Pete and Bill Jaques, who owned houses on Carver's Harbor. Bill Jaques' mother-in-law was a lifelong friend to Cynthia.

In the 1970s, the McAdoos began to rent the Bill Jaques house for a few weeks every summer. Mac remembers the first time they sailed into Carver's Harbor and, as previously instructed, picked up a Jaques mooring. Out came Peter Jaques, who hadn't been told they were coming. Peter said, "What the hell are you doing on my mooring?" And so the McAdoos came to Vinalhaven.

II

Steamboats and Smithy Boats

Island Steamboats in Lore and War

Having drunk a flask of Florence [wine] *at a tavern and flung the empty flask on the fire, he called for a basin of water to wash his hands. When a small quantity* [that] *remained in the flask began to boil, it occurred to him to try what effect would be produced by inverting the flask and plunging its mouth in the cold water. Putting on a thick glove to defend his hand from the heat, he plunged its mouth in the water. The liquid immediately rushed up into the flask and filled it. This suggested to him the possibility of giving effect to the atmospheric pressure by creating a vacuum in this manner.*

The events of this convoluted description by the Englishman Thomas Savery occurred in 1698. That same year, Savery obtained a patent for "raising water by the impellent force of fire." Although the idea of steam power dates back to Hero of Alexandria in the first century AD, many feel that Savery deserves credit for putting the principle of condensation into practice.

In North America, it all began with John Fitch, who built the first steamboat in 1785 and launched the first commercial steamboat service in June 1790. Fitch's steamboat was sixty feet long and carried passengers and

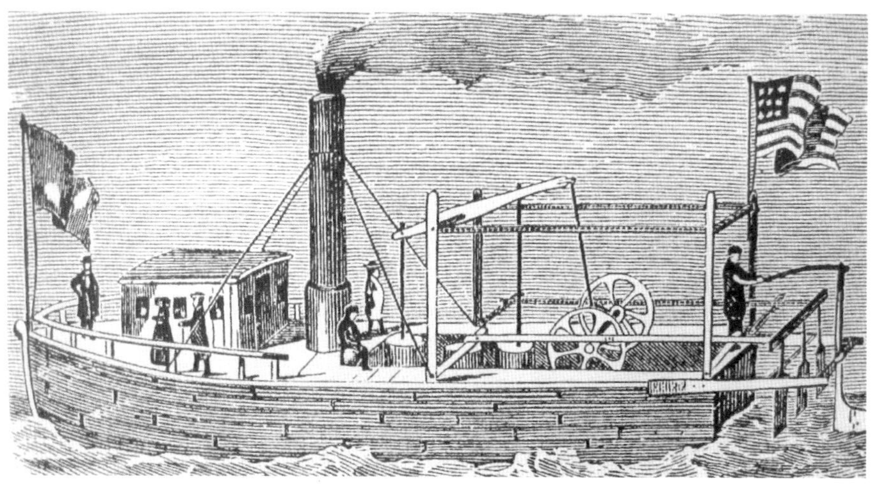

An artist's conception of John Fitch's steamboat, first launched in 1790. *Author's collection.*

freight between Philadelphia and Trenton at the unheard-of speed of eight miles an hour. It made the round trip three days a week and traveled a total of three thousand miles. Although his invention was a technological success, Fitch lost money on every trip, and at the end of the summer his investors abandoned him. Although his business failed, Fitch foresaw what was to come: "The day will come when vessels propelled by steam will cross the ocean! And I almost venture to prophesy that the same power will be utilized in moving vehicles on land."

In the early nineteenth century, an intense competition developed between Americans Robert Fulton and John Stevens, each seeking to be the first to design and run a paddle-wheel steamer. In 1802, Stevens built a screw-driven steamboat, and in 1807, he built an oceangoing steamboat that sailed from New York to the Delaware River. Fulton, however, is generally credited with building the first commercially successful steamboat, *Clermont*, which carried passengers from New York to Albany, a round trip of three hundred miles. A smug Fulton wrote to a friend, "I overtook many sloops and schooners and parted with them as if they had been at anchor and yet when we left New York there were not thirty persons who believed the boat would ever move."

Steamboat transportation grew rapidly, and by 1823, there were three hundred steam-powered vessels operating in American waters. The first appearance of a commercial steamboat in Maine occurred in 1819 when a boat built by Jonathan Morgan carried excursion parties from Bath to

The Steam-Boat

IS now ready to take Paſſengers, and is intended to ſet off from Arch ſtreet Ferry in Philadelphia every *Monday*, *Wedneſday* and *Friday*, for *Burlington*, *Briſtol*, *Bordentown* and *Trenton*, to return on *Tueſdays*, *Thurſdays* and *Saturdays*—Price for Paſſengers, 2/6 to Burlington and Briſtol, 3/9 to Bordentown, 5ſ. to Trenton. June 14. tu.th ſtf

Fitch's announcement. *Author's collection.*

Portland. Captain Seward Porter, however, is considered to be the father of steam navigation in Maine. In 1824, he designed a craft "for local excursions" that had a small engine mounted in the hull of an old flat-bottomed boat. Two years later, Seward opened a route that transported passengers from Portland to Boston and New York. The *Portland Argus* announced that "the steamboat *Patent* is strong and commodious and elegantly fitted for passengers. Her engine has been proven and is of superior workmanship and propels the boat at about ten miles an hour." According to a contemporary source, *Patent*'s engine "could transport a vessel to Russia."

As the number of steam-powered vessels in Maine increased, tensions developed between rival steamers. In 1826, the steam brig *New York* began a schedule with stops in Boston, Portland, Bath, Belfast and Eastport. A fierce competition soon developed between *New York* and *Patent*, which resulted in a number of apparently deliberate collisions. One such incident took place off Owls Head, disabling *Patent*. Having put *Patent* out of operation, *New York* magnanimously towed her eighty-foot rival to Belfast for repairs before continuing on to Eastport.

New York received her comeuppance later that evening. A fire broke out as the vessel was nearing Petit Manan Light. According to a passenger:

> *A glimmering light was discovered around the port funnel. Only two men were on deck: one at the helm and one at the bow. No engineer or fireman was at his post, and but one bucket could be found on deck. Before assistance could be had, the fire had got the upper hand, and the engineer could not stop the machinery. No fire engine, hose or buckets could be found to throw a drop of water.*

Fortunately, the passengers escaped in the lifeboats and landed about midnight at the lighthouse. From there, they were transported to the mainland.

By the 1850s, relatively seaworthy and reasonably safe steamboats were maintaining schedules along the New England coast from Boston to Portland and on to Rockland and Bangor. Transportation to the islands was another matter, which brings us to the so-called Penobscot Bay Steamboat Wars.

The Penobscot Bay Steamboat Wars

The Penobscot Bay Steamboat Wars were a lengthy and spirited competition between rival steamship companies. Each company was seeking to have the fastest boat on the increasingly lucrative ferry route between Rockland and the Fox Islands (Vinalhaven and North Haven). The "wars" unofficially began in 1873, when the Frenchman's Bay Steamboat Company challenged the Fox Island and Rockland Steamboat Company. The names of the boats would change, but the wars would run well into the twentieth century.

In his book *Fish Scales and Stone Chips*, Vinalhaven historian Sidney Winslow tells us that in 1867, the New Hampshire–built *Pioneer* was the first steamboat "to ply a regularly scheduled service in Penobscot Bay. *Pioneer* was the cause of much rejoicing, and people from the remotest parts of the Fox Islands came to get a view of the wonder craft that was to be their communication with the outer world." Faster vessels came and went, but the little ninety-two-foot-long *Pioneer* was to run from 1867 to 1892.

For the next half-century, the Fox Islands would be served by more than twenty different steamboats, even though many were in service for only a short time. Although most boats have long since been forgotten, those with names like *Forest Queen*, *Emmeline*, *Mayfield*, *Golden Rod*, *Sylvia* and *Island Belle* are a reminder of a departed era.

Who Would Be the Fastest?

Pioneer's first challenger was *Ulysses*, which was put on the route by the rival Frenchman's Bay Steamboat Company. In 1873, her owner, Captain David Robinson, saw a favorable business opportunity and decided to challenge

The End of the Packet Era

It should be noted that prior to Pioneer's appearance in 1867, Vinalhaven and other islands in Penobscot Bay were served by sailing packets that, weather permitting, visited the islands every three weeks. (The phrase "packet service" refers to a regularly scheduled line that carried freight and passengers.)

The sailing packet era was a short one, however, since wind-powered vessels were unable to provide reliable service to the islands. With the growth of the granite industry on Vinalhaven, the island needed a more regular and reliable transportation service to the mainland, and with the end of the Civil War, it finally began.

Indeed, "regular service" by sailing packet was something of an oxymoron in the mid-nineteenth century. The service had begun in 1855 with Captain John Carver's little vessel, *Greyhound.* Carver had the ship especially built to serve the route. Two more packets, *Golden Rule* and *Medora*, would follow, the latter skippered by James Arey.

Pioneer's monopoly on the Vinalhaven run. The 239-ton side-wheeler *Ulysses* was more than twice as heavy as *Pioneer* and considerably faster. *Ulysses* had already been in service since 1864 and was a proven contender on the Rockland–Mount Desert route. For the next few months, *Ulysses* dominated the Fox Islands route, forcing *Pioneer* to temporarily withdraw from service.

According to Rockland native John M. Richardson, a formidable rival to *Ulysses* soon appeared in the form of *Clarita Clarita*, a converted fireboat from New York City. Richardson's authoritative book, *Steamboat Lore of the Penobscot*, was published in 1941. He uses a baseball analogy when he describes *Clarita Clarita* as "a pinch hitter brought in to bat for *Pioneer* when *Ulysses* tried to break into the island business." *Clarita Clarita* was 115 feet long and extremely fast.

Passengers loved *Clara Clarita*'s speed and her fare—just twenty-five cents round trip—but the company was unhappy with the amount of coal she consumed, as well as with her deep draft. Sidney Winslow wrote, "Her time for crossing the bay from wharf to wharf has never been beaten by any steamer employed on the line." To continue with the baseball analogy:

Pioneer was the first steamboat to provide regularly scheduled service in Penobscot Bay. *Courtesy Vinalhaven Historical Society.*

"After pinch-hitting a single, she was thrown out stealing second." The costly *Clarita Clarita* would eventually be withdrawn from service, but not before forcing *Ulysses* back to her previous Rockland–Mount Desert run.

Ulysses' untimely demise deserves mention. Early in 1878, during a savage winter storm, the side-wheeler broke loose from her mooring and ran onto the ledges at the southern end of Rockland Harbor. Her hull was quickly reduced to splinters and her engine wrecked beyond salvage. Like most ships of the day, *Ulysses* was uninsured, and the Frenchman's Bay Steamboat Company suffered a $200,000 loss.

With the "benching" of the costly *Clara Clarita* and the exile and later destruction of *Ulysses*, a refurbished *Pioneer*, captained by William R. Creed, was put back on the Vinalhaven route. After twenty years of faithful service, however, it was becoming clear that *Pioneer* was simply not equal to the task of serving the burgeoning Fox Islands population. One problem: she was too slow. Reportedly, her time for the fifteen-mile crossing ranged from two to five hours. Jokes abounded, according to Sidney Winslow: "Considering the amount of time it takes, the trip to Rockland on *Pioneer* is not expensive. For seventy five cents you can stay on the boat all day."

In the fall of 1891, a number of dissatisfied Vinalhaven residents met with George Kimball, president of rival Frenchman's Bay Steamboat Company. With the growth of the granite industry on Vinalhaven, the island's population had expanded rapidly, from a few hundred people to almost three thousand by 1880. Clearly, a faster and more reliable transportation service to and from the mainland was needed. Town merchants were also unhappy. Because of cheap fares, more and more island folks were doing their shopping in Rockland, to the detriment of Vinalhaven's shopkeepers.

During the meeting, company president Kimball announced that he had purchased the steamer *Emmeline* and that she would shortly be put into service to oust the aging and slower *Pioneer*. Reassured by his promise, many citizens agreed to give Kimball's new outfit, the Vinalhaven Steamboat Company—a subsidiary of the Frenchman's Bay Line—their business. At about the same time, W.S. White, manager of the Fox Island and Rockland Steamboat Company (aka the Rockland Company), announced that he had leased a new steamer, *Forest Queen*, and was prepared to challenge *Emmeline* for supremacy on the route.

Emmeline *v.* Forest Queen

The year 1892 would be a momentous one in the annals of island ferryboat competition. Partisans for each boat split into warring camps, and the Penobscot Bay Steamboat War broke out in earnest. The question of which line had the faster boat provoked intense wagering and fistfights in Rockland bars and pool halls, resulting in frequent calls for the police.

Vinalhaven was literally split into two camps. The majority supported *Emmeline*, although a significant minority favored W.S. White's entry, *Forest Queen*, the boat he had chartered to replace *Pioneer*. *Forest Queen* was the first steamboat in the area to have electricity. In addition, she combined speed and comfortable accommodations for passengers and crew. *Emmeline*, the so-called People's Boat, was also a handsome craft, considered more seaworthy, and one that inspired intense loyalty. According to one wag, "Her supporters would rather row than take *Forest Queen*."

Emmeline was the first to go into service in the spring of 1892, and she quickly proved to be a much faster boat than the venerable *Pioneer*. Although many islanders remained loyal to the elderly vessel, it was apparent that her days as a ferry were numbered. When *Forest Queen* arrived on the scene, *Pioneer* was quickly retired and sold to the Maine Central Railroad for use as a lumber lighter.

In April 1892, the time had come to see which of the two new boats, *Emmeline* or *Forest Queen*, was the faster. A clipping in the *Rockland Courier-Gazette* described the atmosphere as the two boats prepared to square off: "Monday and Tuesday of last week were days of excitement such as Vinalhaven has not had for a long time. On Monday, when the hour approached for the boats to show up on the trip from Rockland, there were crowds lining the shore. Friends of each boat were confident that their favorite would be the victor."

Sidney Winslow picks up the story from there: "All eyes were directed toward Green's Island Point, when all of a sudden the tall smokestack of *Forest Queen* shot by the point and came to finish first." *Emmeline* arrived four minutes later, spreading gloom among her supporters. The next day, however, the order of finish was reversed, and *Emmeline* finished first.

Winslow, although a partisan of *Emmeline,* eventually admitted that *Forest Queen* was the faster boat. *Emmeline*'s victories usually occurred when she carried fewer passengers and little freight. The careers of both boats, however, were short-lived. On May 1, 1892, *Emmeline* hit a ledge in the fog off Green's Island and had to be towed off—ironically, by *Forest Queen*. She stayed on the route until 1892, however, when another newer and faster steamer, *Vinal Haven*, was built to replace her.

Forest Queen's problem was that she did not handle well when the sea was rough. It was reported that many of her passengers "fed the fishes" when the wind blew up and she began to roll. In June 1892, *Forest Queen* was removed from service, and the Rockland Company replaced her with *Governor Bodwell. Forest Queen* was returned to Portland and eventually sold to shipping interests in Cuba.

The next round of the steamboat wars began in the summer of 1892. The Vinalhaven Steamboat Company (aka the Frenchman's Bay Company) was building the aforementioned *Vinal Haven* to challenge the Rockland Company's new entry, *Governor Bodwell.* While *Emmeline* was being repaired after her encounter with the ledge, a fast little steamer, *Viking*, was briefly brought in to challenge the Rockland Company's *Forest Queen*. There was a single race between *Viking* and *Forest Queen* that appeared to favor the former, until a "puff of steam enveloped *Viking* and she began to slow down and finally stopped," done in by a broken steam pipe. Winslow continues with baseball lingo when he tells us "the Mighty Casey has struck out."

By July 1892, both of the new steamers were online. The Rockland Company's *Governor Bodwell* took over from *Forest Queen*, which had been leased from the Casco Bay Line in Portland, and *Emmeline* was replaced

The Rockland Company's *Governor Bodwell* took over from *Forest Queen*. *Courtesy Vinalhaven Historical Society.*

by *Vinal Haven*, built in Searsport, Maine. *Vinal Haven* was eighty-six feet long and carried with her the hopes of most of the Vinalhaven community. Even the town band made the trip to Searsport to provide her with a return escort.

Governor Bodwell *Triumphant*

Much to everyone's disappointment, *Vinal Haven*'s appearance and performance were less than impressive. When Sidney Winslow got a glimpse of their new champion, he said, "I swallowed—and swallowed hard." The ungainly pilothouse became the butt of jokes. "It looked like a packing box. There was nothing of dignity in her entire anatomy, with the possible exception of her hull." Unfortunately, her engine was already outdated, and the skinny smokestack had been painted an unsightly yellow. What had been an even contest now shifted to favor the Rockland Company's stylish new boat, *Governor Bodwell*. Everyone agreed: had a race been held that summer, *Vinal Haven* would have been disgraced.

What must be considered a blessing in disguise occurred early on the morning of January 13, 1893. A fire broke out in *Vinal Haven*'s coal bunker, and the superstructure of the wooden ship was badly damaged. Fortunately, what remained of the boat was scuttled at her berth in Rockland before the fire reached the hull. John M. Richardson wrote, "If it were of incendiary origin it must have been set by one of her friends. Her enemies were entirely satisfied with the status quo." Talk about being damned with faint praise.

The *Courier-Gazette* reported that when the fire broke out, "*Governor Bodwell* lay at the wharf and her crew was the first to arrive and render assistance." Happily, the crew of *Vinal Haven* escaped, although Arthur Mills lost everything and was severely burned before he got out. Mills received sympathy in the form of this verse from a local wit:

> *"Bullet" Mills was in his bunk*
> *He got out before she sunk.*

Although *Vinal Haven* was not insured, the hull was raised and her engine cleaned up. Four days later, the boat was on her way to Searsport—even more remarkably, under her own power. A strike delayed the rebuilding process so that it was six months before a redesigned and refurbished *Vinal Haven* was back in the water. A new engine was installed and a more attractive and functional pilothouse and passenger cabin added. Although *Vinal Haven*'s engine was vastly improved, *Governor Bodell* was still considered the faster boat.

Naturally, there were still those who yearned for a race between the remodeled *Vinal Haven* and *Governor Bodwell.* Sidney Winslow describes an informal contest that occurred when Hunting's Circus visited Rockland. Both boats took excursion parties to Rockland in the morning. On the return trip, *Vinal Haven*, being the first to fill with passengers, got a ten-minute head start on her adversary. Halfway across the bay, *Governor Bodwell* began to catch up. Then she began to slow down and finally stopped altogether. After a few minutes' delay, she started up again. *Vinal Haven*, however, was the first to reach Carver's Harbor. *Governor Bodwell* supporters claimed that their boat would have won were it not for an overheated shaft bearing and a brief stop at Hurricane Island to drop off passengers.

Perhaps the beginning of the twentieth century is as good a time as any to declare an end to the ferryboat wars, since following this last "race," both sides felt they could claim victory. As George Kimball, president of the Vinalhaven Steamboat Company, wrote in an open letter, "Our interests are identical in the work in which we are so enthusiastically engaged. May the present harmonious and friendly relations long continue and more firmly unite us in the days that are to come."

Postwar Lore

Over the years, the hard feelings between supporters of the two steamboat lines gradually subsided. In 1898, the Vinalhaven Steamboat Company went bankrupt, and *Vinal Haven* was sold to W.S. White, president of the Rockland Line. For the next thirty years, there was plenty of business for both *Governor Bodwell* and *Vinal Haven* to ferry travelers and freight across Penobscot Bay to Hurricane Island, Vinalhaven and North Haven and as far east as Stonington and Swan's Island. In 1920, *Governor Bodwell*, being the faster boat, was transferred to the longer Rockland–Swan's Island run, whereas the remodeled and larger *Vinal Haven* made the shorter trip to the Fox Islands.

Both boats encountered misfortune in the years that followed. In 1905, *Vinal Haven* got caught in the ice and sank at Tillson's Wharf in Rockland. She was removed from service, given a refit—including an additional fifteen feet in length—with the result that she was "a remarkably well-arranged boat with ample passenger quarters and good freight capacity," according to John M. Richardson.

In January 1924, *Governor Bodwell* ran onto Spindle Ledge, near Swan's Island, during a blinding snowstorm. Local fishermen responded to distress signals, and the *Rockland Courier-Gazette* reported, "Passengers were taken off and the mail saved, though the craft appeared to be a total loss." A month later, after "days of patient, painstaking effort" by Captain John I. Snow, the ship was brought to the surface and the damage found to be less than initially feared. A decision was made to rebuild and modernize "the faithful old craft," which was mostly covered by a $40,000 insurance policy. After a refit, she was put back into service.

After years of continued service, *Governor Bodwell*'s existence came to an end in March 1931 when she caught fire at her Swan's Island berth and was completely destroyed. John M. Richardson wrote, "In a matter of minutes the harbor was as bright as day from the devouring flames. Gathering islanders stood hopelessly by, as lines were cast off and the fiercely blazing craft towed clear of the threatened wharf by a motorboat." Richardson adds, "No Penobscot Bay steamboat has

Governor Bodwell ran on Spindle Ledge off Swan's Island in 1924. *Courtesy Vinalhaven Historical Society.*

ever received the genuine affection which was universally accorded the *Governor Bodwell*." Could there be a better epitaph?

By 1938, *Vinal Haven* had become a "spare" boat used only when her replacement, *W.S. White*, was being repaired. *Vinal Haven*'s demise occurred

W.S. White was requisitioned by the navy at the start of World War II. *Courtesy Vinalhaven Historical Society.*

in November of that year when, according to a *Portland Press Herald* story, "she caught on a piling at Tillson's Wharf and began to list heavily. With the incoming tide she remained partially submerged despite efforts to right her." *Vinal Haven* was condemned, stripped of everything valuable and abandoned at one of the old "Point Kilns" docks. Finally, in 1945, her hull was towed out of Rockland Harbor and sunk between Munro and Sheep Islands, beyond Owls Head.

The United States entered World War II in December 1941, and the nation immediately went on wartime footing. *W.S. White* was requisitioned by the navy and sent to the Caribbean, where she served as a dispatch boat. Shortly after this, company president White announced that after seventy-five years, the Rockland-Vinalhaven Steamboat Company would be going out of business.

For over a year, the Penobscot Bay islands had no regular ferry service. Local fishermen and boat owners filled in as best they could, carrying passengers and freight back and forth to the mainland. Then, at a special town meeting in August 1942, Vinalhaven voted to raise $55,000 to build a powerboat. The result was a sixty-five-foot, diesel-powered "motorship"

The sixty-five-foot *Vinalhaven II* was able to carry two cars. *Courtesy Vinalhaven Historical Society.*

named *Vinalhaven II*, built in Southwest Harbor, Maine. The boat went into service in July 1943, and Charles Philbrook was her captain—but that begins another story.

In Search of the Smithy Boat: A Maritime Detective Story

The name sounds like a World War II German E boat or a coastal raider from World War I. In reality, the Smithy boat story is a bit less exciting, although its origins are equally obscure. Smithy boats first appeared on Vinalhaven in the late nineteenth and early twentieth centuries, when it was determined that gasoline engines could be used to power fishing boats as well as pleasure craft. How extensively they were used elsewhere on the coast is hard to determine.

If one looked at a picture of Carver's Harbor on Vinalhaven one hundred years ago, one would see a lot of sailboats and a few funny-looking little craft that looked like sloops but lacked a main mast. These were the Smithy boats. They came in different sizes, but they had a few features in common. One

Smithy boats line the shore of Creeds Cove on Vinalhaven, circa 1920s. *Courtesy Vinalhaven Historical Society.*

characteristic was an undercut stern. Another was a large, four- or five-foot keel; a few had a small stern sail. All of them had a small one-cylinder—or "one-lunger"—engine that sat in the center of the boat, or at the stern, and was exposed to the elements. Some of the boats had a low spray hood. Later designs also had a small cabin called a "cuddy."

Smithy boats were clearly influenced by the design of the Friendship sloop. They might have been built as early as the late 1890s and were probably built for about thirty years, until the late 1920s. There was also considerable variation in their size and design. Phil Dyer, who has built over one hundred boats himself, said, "To build a Smithy boat, you have to know a lot about boat building because of the way the stern and keel are constructed."

Before the advent of the gasoline engine, oars or sails powered fishing boats. The *dory* was a high-sided flat-bottomed rowboat used by many New England fishermen beginning in the late eighteenth century. The *peapod* was named for its double-ended, round-sided shape resembling the garden vegetable. It was thought to have originated on Penobscot Bay in the mid-nineteenth century. The *Friendship sloop* began its career as a fishing and lobster vessel in Muscongus Bay. Only later did it evolve into the familiar pleasure craft we know today.

According to Vinalhaven historian Roger Young, once gasoline engines became reliable, fishermen realized they needed a different kind of boat. Boats with engines could extend the range, as well as the season, for fishermen. Thus, the Smithy boat was born as an early powerboat. As noted, the Smithy boat hull was similar in design to that of a sloop, with a large keel and undercut stern. Instead of a main mast, it often carried a small stern "jigger" sail for stability in heavy weather. Over time, the design was modified. In addition to the "cuddy," the keel length was reduced so that fishermen could set their traps closer to the rocks and ledges.

Possible Origins

In search of information about the Smithy boat, I spoke with a number of retired Vinalhaven fishermen and boat builders who were familiar with the design. Some of them had owned Smithy boats when they were young men. Others remembered them from the days when their fathers and grandfathers owned them.

There are several possibilities as to the origins of the Smithy boat. In the early twentieth century, there was a Vinalhaven boat builder named Ernest H. Smith; an advertisement for his boats appears in the Knox County section of the 1907–08 Annual Maine Registry of Businesses. Further investigation reveals that Smith was born on Vinalhaven in 1875 and died in 1944 in Brookline, Maine. Certainly, Ernest Smith is a possible Smithy boat builder.

Then there is Hibbert Smith. Records in the Vinalhaven Historical Society indicate that he was born in Nova Scotia in 1862, moved to Vinalhaven as an adult and died on the island in 1929 at the age of sixty-seven. Smith's occupation was listed as "fisherman." Did he also build boats during the winter for extra income? There are those who think he did.

Ivan Olson is one of them. Olson is a retired lobsterman in his eighties who fished out of Vinalhaven his entire career except for a stint in the navy during World War II. Olson remembers hearing about a Hibbert Smith who built boats down on Clamshell Alley. Clearly, the fisherman Hibbert Smith is a serious candidate.

A third option is that the Smithy boat was not named after the *builder* but after one of the first *owners*. Vinalhaven resident Doug Hall remembers when Smithy boats were built in Carroll Gregory's boat shop just off Main Street on Clamshell Alley, the same shop that Hibbert Smith formerly used. In fact, Gregory built a Smithy boat for Doug's grandfather in 1917. Is it possible that the first Smithy boat was built in Carroll Gregory's boat shop for a Vinalhaven fisherman named Smith early in the twentieth century?

Vinalhaven resident and local historian Roger Young suggests that although Smithy boats may have first appeared on Vinalhaven, the name became a generic term for the hybrid fishing craft that was used in the early twentieth century on the Maine coast. I have checked photos of coastal harbors from one hundred years ago and found a few craft that looked like Smithy boats, though not as many as were in Carver's Harbor on Vinalhaven.

For example, there is a picture (taken in 1907) of a Smithy boat in Dean Lunt's book *Hauling by Hand*, a history of the town of Frenchboro. Retired North Haven boat builder Bud Thayer told me that his father had a Smithy boat and that he remembers repairing some of them, though he associates them mostly with Vinalhaven. Paul Stubing from Deer Isle and Paul Pendleton from Islesboro also remember seeing local fishermen using Smithy boats but agree that they originated on Vinalhaven.

Ivan Olson began lobstering at the age of ten when a couple of "old guys" gave him some traps to patch up. He remembers that Smithy boats were in use when he was a kid, and as noted, he thinks Hibbert Smith built them in his Clamshell Alley shop. Ivan told me, "They were popular in the 1920s. There were a lot around Vinalhaven in those days, but I never saw them anywhere else. Today they are a forgotten boat."

Ivan recalled, "Although I never saw any on coast, some may have migrated to North Haven from Vinalhaven. They were widespread in Vinalhaven and were used exclusively for fishing and lobstering, but never for pleasure. No one had money for pleasure boats in those days." Olson doesn't know why they had such big keel. "Probably it was the influence of sail boats. Some had a wheel; some had a tiller. Most boats used a Ford Model T engine, converted to a marine engine, although there was a lot of variation in the design." I asked Ivan why Smithy boats disappeared, and he shrugged, "I guess guys just wanted bigger boats."

Looked Like Pumpkin Seeds

Ivan described the typical Smithy boat as being "maybe fifteen to twenty feet long; none of them were very big. They looked like pumpkin seeds," he added. I asked Ivan how far out to sea he went, and he told me three or four miles, adding, "There were plenty of fish to go around." Ivan told me, "Matinicus fishermen claimed the waters around Seal Island and Wooden Ball in those days. I stayed away from Matinicus, which was just like it is now. Once in a while we even had fights with them."

Dean Lunt confirms the limited range of the Maine fisherman in *Hauling by Hand*. "Fishermen might row or sail a couple of miles or so each way to haul traps which numbered maybe fifty to seventy traps. In fact, some of the older fishermen continued to row around the island [Long Island] hauling their traps by hand into the 1950s."

Ivan Olson emphasized that the Smithy boat was built especially for fishing and was not just a modified Friendship sloop. "They were good sea boats and were comfortable and quite rugged," he said. "Better than the stuff they are building today," which he calls "Tupperware boats." Olson fished for pollock, cod and hake and did some lobstering. "In those days, I'd be embarrassed to say I lobstered," he told me. Vinalhaven's Ivan Calderwood wrote in *Uncle Dave's Memoirs*, "Trawler fishermen sometimes laughed at the little lobster catcher and called him 'a landlubber,' as they sailed by in their big trawlers headed for the Grand Banks to fish."

Vinalhaven fisherman and lobsterman Dallas Anthony is in his seventies and remembers when his father, Francis, had a Smithy boat. Francis Anthony was born in 1908, and Dallas told me he bought his boat from a guy on North Haven. "When I was a kid they were all over the place." He showed me a picture of his father taking some of the family for a Sunday afternoon outing on *Edith A* that was probably taken in the 1930s. Dallas confirmed that Smithy boats were fishing boats, even though "they looked like sloops." He also told me that a lot of the old boats had no names.

Doug Hall was born in 1929 and grew up on Vinalhaven. He is a Korean War veteran and a retired German professor from the University of Maine. Doug began fishing in 1943 and has owned two Smithy boats. He bought his first, a little twenty-one-footer, in 1948 from Josh Williams for seventy-five dollars. "Josh needed a larger boat for fishing."

Doug told me that when he got the boat "the engine was shot, but that his friend Harry Philbrook somehow got it going." The stern was falling apart, but legendary Vinalhaven boat builder Gus Skoog replaced the

Duck hunting on a Smithy boat, off Leadbetter Island, circa 1912. *Courtesy Vinalhaven Historical Society.*

Ivan Olson, ship's cook third class, 1944. *Courtesy Vinalhaven Historical Society.*

sternpost and planks for $150. Doug then put in a Model A engine and a new transmission. He soon realized the boat wasn't large enough to use for haking, so he sold it to Herb Peterson, who used it for lobstering for the next ten years.

Doug bought his second Smithy boat in 1949 from Pete Dyer. "Pete had made a killing on herring," he told me, "and he decided to buy a larger boat." Doug's new boat was twenty-six feet long by nine feet wide, and it drew four and a half feet of water. He told me Carroll Gregory had laid keel in 1917, and it was still "absolutely watertight when he used it thirty years later." Pete Dyer had built a canopy and a cuddy cabin with a stove. Doug then added a new engine and a gaff-rigged sail on the stern.

With his "new" boat, which he named *Flash,* twenty-one-year-old Doug Hall was ready to go haking. Doug remembers leaving Carver's Harbor in the middle of the night with his six-cylinder Chevy engine wide open. He told me he'd never run a boat at night before and that he panicked when he hit a fog bank that was so thick he couldn't see the bow of the boat. He turned *Flash* around and slowly crept back into harbor.

As he entered the harbor, Doug had the good fortune to run into his brother David, who couldn't get his boat engine started. The two brothers decided to combine forces and go haking together in Doug's new boat. "We made a real killing," Doug told me. "In fact, David always said it was best boat he'd ever been in for hauling hake." (According to Ivan Olson, what made Smithy boats so good for hauling hake was that the boat was beamy and had lots of storage space.)

Doug had some close calls on *Flash.* One time, off Wooden Ball Island, south of Vinalhaven, he was trying to find his gear (nets) in a thick fog, and a big sea came rolling in. He headed for the lighthouse at Matinicus Rock by listening for the foghorn, which helped him get oriented. Doug fished through the summer of 1950 before he went to Korea, "when most of the shooting was over."

I asked Doug how seaworthy his Smithy boat was, and he told me, "She was very stable riding with the 'jigger' sail, which steadied the boat. In rough weather, the prop sometimes lifted out of water, but if I went slowly, while

I was hauling gear, I could adjust the sail as I pulled up the nets. The one-lungers gave it enough power, and you could go forever on a gallon of gas."

Doug told me he found that haking was much more interesting than lobstering. "Each day was different, depending on what you found in the nets. The tide was always a factor, as was the number of fish. You could have one thousand pounds of fish and gear coming up from forty-five to fifty-five fathoms, including dogfish and skates. Every day was just an adventure." When he went into the service, Doug sold *Flash* to Frank Adams, principal of Vinalhaven High School, who lost it shortly afterward. "I heard it got loose from the mooring and broke up when it went ashore."

"An Able Son of a Gun"

Bert Dyer graduated from Vinalhaven High School in 1940. That summer, he bought a Smithy boat from his brother Les for $150. It was twenty-six feet long and had a four-cylinder Gray engine with a rounded "steamboat" stern. (The steamboat stern indicates that Bert's Smithy boat was of a later design.) "It was a good boat, well rigged and fresh-water cooled," said Bert. "It never gave me any trouble. It had a small prop, a pot hauler and a spray hood. She was nice and wide and comfortable as hell. I could go to Rockland and back for seventy or eighty cents worth of gas. I used her for hauling lobster and haking and trawling outside of Matinicus Rock, and I could run her by myself. Once I brought in $4,000 worth of fish in four tubs. What an able son of a gun she was."

Bert didn't know why they were called Smithy boats. All he knew was that they were built on Vinalhaven and that there was a lot of variety in them, which supports what others have said. The longest were probably thirty to thirty-two feet. Most were twenty-six feet or less, although they were all very seaworthy. "Some you had to shut the engine off when you stopped; some you didn't. Some had a mast; some didn't." Bert told me his boat had a jigger sail on the stern for stability. "That would steady them down if was blowing."

Bert didn't recall seeing Smithy boats on North Haven or Matinicus, although "there might have been a couple in Rockland." He had his boat for a year, then sold it for $125 and went into the navy in 1942. Bert told me the guy who bought his boat "let the engine freeze up." He later sold it to a guy who lived in Dogtown on Old Harbor.

Bert told me the tragic story of Carl Magnuson, who owned a Smithy boat but his engine broke down. While he was waiting for a new engine, he went fishing in his dad's Novi boat. In June 1937, they were off Seal Island, near

Matinicus, and Carl slipped off the boat while pulling in a line and got tangled in his sweater. His father jumped in after him, but Carl, who couldn't swim and had his boots on, sank before he could be pulled out. Carl's body was never found.

Although he never owned one himself, Bert's brother Phil Dyer remembers the Smithy boat his grandfather bought after World War I. It was eighteen feet long, and it had a small four-cylinder engine that was very reliable. His grandfather used it for trawling, fishing and lobstering. One of the highlights of Phil's boyhood was the time he took the boat to Rockland "to get some booze." Phil remembers seeing fifteen to twenty Smithy boats in the waters around Vinalhaven when he was a kid and that most of them were "real deep," drawing five to six feet.

The End of an Era

Smithy boats were like old cars: they had lots of owners in hard times. They were popular during the Depression in the 1930s because they were cheap to operate. A lot of older fishermen who owned Smithy boats were reluctant to change over to the new marine engines and continued to use the original one-lunger engines, or old car engines.

The next generation, however, wanted something better. After World War II, Smithy boats began to disappear and were replaced by the more efficient modern lobster boats. The deteriorating Smithy boat hulls were pulled up on shore, where they remained for a long time. The Vinalhaven Historical Society has a picture, taken in the 1950s, of what may be Bert's Dyer's old boat with its rounded stern, covered with snow, sitting on the shore at Old Harbor.

Up to the 1960s, people continued to congregate in Carroll Gregory's boat shop to watch him work and listen to his stories about the "old days." Phil Dyer learned the craft of boat building from Carroll Gregory and Dick Young, who at that point was in his eighties. Phil said he was told "to sit quietly and watch how boats were built." He remembers that Dick Young kept a bottle of Light House rum hidden in shavings in a nail keg. Once in a while Dick would reach in and take a drink. "I have throat trouble, and this is my medicine," he told young Phil. "Just don't tell my wife." The "medicine" must have worked because Dick lived to be ninety-four!

We may not know the exact origins of the Smithy boat, but we do know that Carroll Gregory, probably the last of Vinalhaven's Smithy boat builders, died in 1966 at the age of sixty-nine. I wonder what he would have to say about the "Tupperware boats" lobstermen are using today.

III

Sloops and Schooners

The Professor Builds a Sloop

Just as of houses, so it is said of yachts,
that fools build boats for wiser men to sail them.
*—Conor O'Brien**

Forty-five years ago, Peter Sellers was a young mathematics professor from Philadelphia with a dream. Specifically, he wanted to build a boat with the lines of a Friendship sloop. The project would combine two of Peter's favorite activities. As a child, he had developed a love of woodworking, a skill nurtured by his father. As a youth, he loved the small boat sailing he had done during summers spent on the New Jersey shore.

When Peter's father died unexpectedly at the age of seventy in 1971, Peter wrote in his journal:

> *His influence over the boat building is all I will mention here. He taught me woodworking and delighted in every project I did. He discussed every aspect of the boat-building project with me and, if he had lived the many more years*

* Edward Conor Marshall O'Brien was a quirky Irish intellectual and a pioneer of modern maritime theory whom Sellers admired and quotes frequently in his logs. O'Brien was also a shipbuilder/designer and, during the Irish war for independence, a gunrunner. He and three friends sailed around the world in a forty-two-foot yacht between 1923 and 1925.

Young Peter Sellers. *Courtesy of the Sellers family.*

we expected, he would have done all of the more delicate and artistic wood carving which is to go in the boat. As Nicholas [Peter's brother] *said, "If you leave a piece of wood in his living room, you'll find it carved when you come back."*

Peter found an eight-inch Friendship sloop in his father's shop carved in relief, which was to have been cast and given to him for Christmas. At the same time, his mother asked that Peter take his father's workbench and tools, the oldest of which had been handed down from his great-grandfather, who died in 1906.

Getting Started on the Dream

By 1970, Sellers was established enough professionally to get started on his dream. With a wife and four young children, however, he realized that he was in for a long-term project. Sellers was nothing if not sanguine, realizing that he would only be able to work on weekends and even then only during the winter, since he and his family spent their summers on Mt. Desert Island. Peter, however, was and still is a patient man. "It would take as long as it had to take," he observed.

Although he loved sailing and enjoyed woodworking, Sellers admitted that he knew little about boat building, especially constructing a craft as large as the one he had in mind. He was, however, familiar with Howard Chapelle's book *Boatbuilding*, which gave him the general principles to follow. "It became my Bible throughout the process," he recalled.

Peter's first step was to get in touch with the publisher of *Boatbuilding* at W.W. Norton & Co. He requested a set of plans Chapelle had made for one of the boats he wrote about in another book, *American Small Sailing Craft*. The plans were based on the old gaff-rigged workboat that plied the waters of Maine in the early twentieth century. Sellers proceeded to scale the plans up 54 percent (remember he is a mathematics professor) for a sloop that would be thirty-eight feet long by eleven and a half feet wide. This would just fit into the barn on his farm in Doylestown, Pennsylvania, where he intended to build the hull. (It took him the better part of a year to renovate the barn and assemble the plans and building materials, so it was 1971 before actual construction could begin.)

Howard Chapelle was a distinguished naval architect and curator of maritime history at the Smithsonian Institution. In *American Small Sailing Craft*, he described the Friendship sloop:

> *The sloops built on Muscongus Bay in Maine are best known as "Friendship sloops" or "Morse sloops." These names imply a deep keel sloop, having a clipper bow, a counter stern and a strong sheer. The names come from the town of Friendship, where many of the boats were built, and from the Morse family, of this vicinity, whose boats were noted for their excellent sailing and working qualities.*

Early Friendship sloop. *Courtesy of the Sellers family.*

Peter's initial challenge was to find a piece of twenty-five-foot white oak to serve as the keel. He found one at a sawmill in Doylestown, Pennsylvania, but it wasn't properly seasoned. His solution, literally, was to "pickle it." "I packed it in rock salt and built a box around it. I left it in there for a year, and you could see the water dripping out of the box." This became the backbone of the boat, and construction began. Following the guidelines in Chapelle's book *Boatbuilding*, Sellers fashioned the ribs. "I made lots of small ones (called canoe framing), which were easier to steam and bend." When the ribs were in place, Sellers put African mahogany planks over the oak framework.

"Building a boat from scratch was a painstakingly slow process," Sellers recalled. "There were lots of steps, and I made lots of mistakes. Every day I had a new problem, but then I had lots of time." For example, in 1975, he built a new transom (the piece across the stern) after deciding that the original wasn't a good fit. And then there was the day in February 1980 when he opened the barn door and found to his horror that most of his tools had been stolen.

I asked Peter what he used for fasteners, and he told me he used bronze screws, which he then counter-sank and covered with wooden plugs. "My father-in-law told me he was 'too old to help,' but he'd pay for the screws. There were literally thousands, so this was a considerable expense." The result, however, was a boat that is so tightly put together that it has never taken water.

Peter showed me his notebooks. There were twelve, one for each year it took him to build the boat. Each notebook is a log of the work he did on the boat from 1971 to 1982. Included are the challenges he faced in a given year, musings on his work, newspaper clippings, poems, letters, references to visits by friends and thank-you letters from visiting schoolchildren. Scattered throughout are pithy quotations from Conor O'Brien. There are also numerous photos of the boat under construction, as well as pictures and postcards from family and friends. The result is a fascinating journal/diary/scrapbook of the project from start to finish.

Some Notebook Musings

> *March 29, 1971: The boat shop* [Seller's renovated barn] *is already complete, and the boat plans have been drawn full scale in a lofting area, which occupies one third of the shop.*

May 9, 1971: Conor O'Brien said, "There is no more fascinating trade than that of a shipwright. While others regulate their work with a square and a straight edge, his rules are a springy batten and an artist's eye."

March 29, 1972: Walking along Second Avenue I saw some shipwright's tools in an antique store. I bought a caulking hammer, a hammer for drifts, a broad axe, and an adze with a flat blade.

November 25, 1972: Today Rudy Fields [a builder/woodworker] *came. He says that parts of my stem and sternpost are hickory, which were sold to me as oak. The main thing wrong with hickory is its being subject to dry rot, although there are precautions which can be taken against that.*

December 3, 1972: Lucy Bell [his wife] *and I dug a pond, fifteen feet across, to put the lumber for the frames in, when it comes. This lumber is new cut white oak, and if kept moist, it will steam bend better (they say). This is not the first time that digging mud has been an essential part of boatbuilding, and hard work too.*

February 28, 1973: In the wee hours this morning I felt the house vibrate and Sean the dog leapt up in excitement. There had been a six second earthquake at 3:20 AM. It remains to be seen whether the boat shop settled or moved to such a degree as to throw off the position of the boat. It would not take much to throw the long plumb lines visibly off their marks [it didn't].

September 21, 1973: It has been three months [away in Maine for summer] *since my last day in the shop, but the appearance of everything was so unchanged, that it might have been the very next morning, and all my summer a dream.*

March 29, 1974: According to Conor O'Brien a boat looks best when the frame has been set up and the planking just begun. You can see all the graceful curves of her timbers, each differing slightly from its neighbors, but so closely related to them as to carry the eye easily over the intervening space, so distinct as to give a sense of form and perspective hidden by the smooth skin of the finished hull.

March 18, 1975: The children's [Easter] *vacation begins. Obviously, if you want to build a boat you've got to keep at it, but you can't be fanatical about it.*

Peter Sellers working on ribs. *Courtesy of the Sellers family.*

April 30, 1975: The shop doors stayed open and a warm breeze swept through, carrying all the loose sawdust away.

September 24, 1975: There's bound to be something in the boat which defies all customary practice, a problem which I don't see how to solve in any accepted way. So I try something new and different. The transom is a good place for this. Not only is it safely above the water line, but also it is one of the traditional places for flights of artistic fancy.

October 2, 1975: Kay Finney brought her fourth, fifth and sixth graders from Miquon School out to see a boat being built. Some children were very savvy. One asked me where I would step the mast, and another asked what the sail area would be.

March 29, 1976: Conor O'Brien said, "The next stage is the supreme mystery of building: the shaping and fitting of the plank. Their edges nowhere straight, nowhere parallel, and all along worked with a varying level to make a true joint."

October 24, 1976: Martin Hummel and a friend visited today. He has been enthralled with boat building since he was a child and went around my shop with glee saying he had come for encouragement, but I saw no signs of his needing it. He and his friend left, knowing better than I, that a boat builder does not get anywhere just by talking.

November 13, 1976: Like Conor O'Brien I came to the conclusion that nobody was fool enough to build (or even design) the kind of boat I wanted. I am forced to the belief that very few yachtsmen really love their boats.

February 20, 1977: I put in the 3,546th screw. Slocum says that the operation of putting on the planks is tedious, which I would agree with if I were doing it full time. Working intermittently, as I do, it is always a pleasure to get back to the shop.

October 30, 1977: All seams are noticeably tight. Last winter there were a couple of places where light could be seen through a seam. Is it true that planks are less tight in cold weather?

February 24, 1978: The builder (me) does not always follow the plans of the designer (me).

March 30, 1978: Conor O'Brien said, "The only true yacht cruise consists in sailing continuously along an unfamiliar coast, and putting into every navigable creek as one passes it."

April 18, 1978: The presence of the mast hole alters the whole appearance of the boat, because one imagines the mast. I begin to feel now more as one finishing the building of a boat, than starting it.
May 2, 1978: I got a pair of glasses today for close work. They came in handy for taking the angles for leveling.

May 12, 1978: A warm breeze blew through the barn carrying away the sawdust from the band saw. The same breeze carried the smell of lilacs with it. This is the time of year to build boats.

March 29, 1979: The boat's eighth birthday. It was hard to leave the farm, being our first warm spring day. Daffodils in bud, crocuses in full bloom, robins strutting about. Even the boat could enjoy the spring, as every

Sloop *Lucy Bell* at the barn door, Doylestown, Pennsylvania. *Courtesy of the Sellers family.*

barn window was open, allowing the wind to sweep right through and letting the boat feel that element which brings boats to life.

October 18, 1979: I used Father's old jack plane for smoothing the topsides of the boat. It sharpens better than my new planes. Long sweeping motions seem to work best.

February 24, 1980: Looking for a large chisel to do some final work on the vertical support, I discovered that all my big chisels had been stolen. All the planes except father's jack plane were gone. Father's brass cannon was gone as was the walnut handled pocked knife he made me. The door had been forced open and there were large footprints everywhere. [The police were called and the value of the tools that were stolen was estimated at over $5,000.]

April 5, 1980: Mary [his daughter] *says I ought to get someone to do the caulking for me. Good idea, but I think no one would be willing to use my method. It is too difficult, and the experts would say it was wrong.*

April 27, 1980: I got back to Doylestown at lunchtime today after a visit to Cambridge to see "Murder in the Cathedral," starring Therese Sellers [another daughter].

January 1, 1981: At odd moments between Christmas and New Years I have gotten the quarterdeck boarded over with fifteen pieces of teak.

February 21, 1981: The entire ceiling has to come out again and be trimmed off at the ends, where there is a slight unevenness.

January 17, 1982: It was twelve below zero when we got up this morning in Doylestown and it stayed below zero all day. The heater in the shop couldn't cope which meant I had the greatest difficulty mixing up the Thiokol [synthetic rubber].

October 25, 1982: I called Spencer Lincoln in Maine and told him to be ready for the boat on December 1st. He will tell Joel White at the Brooklin Boat Yard [in Brooklin, Maine].

November 7, 1982: I started packing up the cockpit with lumber and other things to go to Maine with the boat.

Peter Sellers beside the hull of *Lucy Bell. Courtesy of the Sellers family.*

November 21, 1982: This weekend Lucy Bell painted the hull and together we dug a trench for the boat to slide into as it moves out of the shop.

December 1, 1982: We heard Neil Levine's [truck driver] *engine start up at 6:27 so I ran out and invited him in for breakfast. He was back in his truck before 7:00, and on his way. We tooted him goodbye and saw the boat fade away into the morning mist.*
December 7, 1982: Spencer Lincoln reports that the boat is in its cradle at White's yard.

Christmas 1982: No Christmas cards this year, we sent boat pictures. On many of the cards we issued invitations to the launching.

Cruising the Coast

When the boat was finished, it was appropriately named *Lucy Bell* in honor of Peter's wife, who had been his invaluable assistant throughout the project. As noted, Sellers had the hull hauled to the Brooklin Boat Yard, on Eggemoggin Reach in Maine, where the mast was stepped and the sails and rigging fitted. Sloop *Lucy Bell* was launched at the Brooklin Boat Yard at high tide on June 21, 1983, to coincide with the Sellerses' twenty-fifth wedding anniversary.

As Peter noted at the launching, "The hull of *Lucy Bell* is a scaled up modification of the sloop *Pemaquid* which was built about 1914 at Bremen, Maine, by A.K. Carter, to be used as a workboat. The sail plan, by Spencer Lincoln, was inspired by that of the single headsail sloops used in Muscongus Bay in the 1870s and 1880s."

Peter and Lucy Bell Sellers have sailed the waters of Maine from Pemaquid to Pulpit Harbor and from Castine to Cutler since 1983. In keeping with early Friendship sloops, *Lucy Bell* does not have a motor; thus, the Sellerses' route depends on the wind and the tide. They may stop to see friends, although they follow no set schedule, as they seek to voyage free of commitments. Each day around mid-afternoon, they look around for a cove or harbor to spend the night before the wind dies down. Every few years, the Sellerses stop in to see us on Vinalhaven. Maybe we'll see them again this summer?

Oh how I love to float on a gaff-rigged boat,
propelled by the wind and the tide.

Sloop *Lucy Bell* under sail in Somes Sound. *Courtesy of James S. Murphy.*

If the breezes fail and I cannot sail,
I can scull or paddle or glide.

But I bar a motor whose unique odor
Is uncongenial to me,
And it's not true sailing when one is trailing
An oily wake on the sea.

—*Peter Sellers, June 19, 1983*

The Rhodes 19 Story

The Rhodes 19 is a racing craft, it is a family day-sailor and, in rare cases, it has been a cruising craft. The boat's history goes back to the post–World War II years (1946), when it was introduced as an all-purpose, one-design boat.

Following the end of World War II, the Allied Aviation Corporation of Cockeysville, Maryland, converted its facilities from making airplane fuselages to producing a product that could compete in the peacetime economy. Accordingly, the company commissioned renowned yacht designer Philip Rhodes to draft the lines for an inexpensive sailboat that was fun to sail. Rhodes came up with a nineteen-foot center boarder, which was initially called a Hurricane. Unfortunately for Allied Aviation, it generated little interest.

The next year (1947) Palmer Scott, a New Bedford boat builder representing the Southern Massachusetts Yacht Racing Association (SMYRA), purchased a number of Allied's unfinished wooden hulls. After fitting the hulls with keels and adding a cuddy cabin, he produced a fast, virtually unsinkable boat. One place where the SMYRA-class boat caught on was at Edgartown, on Martha's Vineyard, where it soon became the basic racing boat for the Edgartown Yacht Club.

Peter Plumb is an attorney who lives in Portland, Maine. He remembers sailing an early Rhodes as a teenager in the 1950s: "At Edgartown there was a large fleet of them, probably twenty-five." Peter said they were known as SMYRAS in those days. "Nobody called them a Rhodes 19. They were beautifully finished, however, with cuddy shelving, lots of varnish and

The author's brother, Joel Gratwick, in his Rhodes 19, *Sarah G. Author's collection.*

mahogany trim." He told me that if a current owner saw an old SMYRA, he'd say, "Gee, I wish I owned that one."

Disaster struck in September 1954 when Hurricane Carol roared up the East Coast. It hit Edgartown and destroyed the entire SMYRA fleet with the exception of one boat. Peter said they took the surviving hull and made a mold for the first fiberglass version.

At the storm's height, the tide was so high that Plumb's boat caught on a pier as it was being blown out of the harbor. It came down hard on a piling, which went right through its bottom. Peter remembers, "The boat was speared as though it was on a spit. The mast got tangled in the wires of a phone pole and shorted out the whole of Chappaquiddick Island [across from Edgartown] for several days while it sat there." The insurance company declared the Plumb boat a total loss. Their reimbursement allowed his family to purchase another SMYRA, which Peter said they sailed for years.

Peter said the SMYRA was fun to race and that the hull and rig made it a sturdy, seaworthy boat. At various times, he sailed to Cape Cod, the Elizabeth Islands and once almost to Nantucket, until it got too rough. The boat was hard to tip over, although he capsized once in a race. The spinnaker sheet caught on the marker, and he went right over.

Peter looked back:

> *It has been an amazing success story for a boat. The design has really stood the test of time. The ocean off Edgartown is a rough place to sail and you need a boat that can stand up to the sea. Sometimes you got wet, but with a long foredeck and cuddy it also made for a nice day sailing boat.*

Fred Brehob grew up sailing in the sheltered waters of Lake Erie's Presque Isle Bay. He went on to Bowdoin College, where he was on the sailing team, before he was drafted and sent to Korea. When Fred got married, he and his wife honeymooned in New England and settled in Marblehead, Massachusetts, where they have lived ever since. Fred is historian of the Rhodes 19, and not surprisingly, he is a veritable font of information about the boat.

I learned that in 1952, distinguished yachtsman George O'Day formed his own company to build affordable sailboats that could be trailered. (George O'Day was the first American to win both an Olympic gold medal in sailing and the America's Cup.) In 1958, O'Day arranged with Philip Rhodes to use his name to identify the boat. The next year, O'Day officially changed the name from SMYRA to Rhodes 19 and began to market the boat up and down the East Coast.

Fred's first experience with an R-19 was with a friend who was one of the founders of the Marblehead racing fleet. They took turns crewing for each other, and in 1963, he told me, "We cleaned up." Fred bought his own Rhodes in 1967, and in a few years he began to win races on his own. Fred still has the boat, though having passed his eightieth birthday he doesn't race anymore.

The Marblehead fleet, however, has continued to flourish. There are about thirty boats in the class that race all summer—on Thursday evenings and on Saturdays. The Martha's Vineyard fleet is not as active, nor is the one in Freeport, Maine. Fred said that currently there are active R-19 fleets in New Bedford, Boston Harbor, Hingham, Chicago and New Orleans. "There was one in San Francisco, but it has faded out."

The only time Fred went on an overnight was when his ten-year-old son persuaded him to take their boat on a camping trip. Their plan was to sail the twelve miles from Marblehead to Gloucester and spend the night. They got as far as Manchester when the wind died. Fortunately, a kindly fisherman appeared who towed them in to Manchester Harbor, where they spent a mosquito-filled night trying to sleep on the Rhodes. Fred said his son never asked to go camping again.

The Voyage of David Pyles

The Rhodes 19 is not designed for extensive cruising, although there are exceptions to every rule. Dave Pyles is retired and lives on the Eastern Shore of Maryland. When he turned sixty in May 2004, Dave wanted to do something memorable. He decided to sail, by himself, up the Intracoastal Waterway in a small boat. After looking around for just the right craft, he met Dave Whittier, the owner of Stuart Marine, at a boat show and bought a fixed keel Rhodes 19.

Dave had the boat fitted for oars and an eight-horsepower Yamaha mounted on the stern. The boat was shipped to the Hinkley yard in Stuart, Florida, where the mast was stepped. Stuart is about forty miles north of West Palm Beach at mile marker one thousand on the Intracoastal Waterway.

Dave sailed his R-19 a thousand miles up the coast to Norfolk, Virginia. The trip took twenty-two days, and he averaged forty-five to fifty miles a day. Dave told me he slept on board the boat more than half the time. (He admitted that sleeping under the cuddy was a bit of a squeeze.) The

Dave Pyles on the Intracoastal Waterway, Belhaven, North Carolina. *Courtesy of David Pyles.*

other nights he went ashore to replenish his food supply, clean up and get a good night's sleep. Most mornings, he left as the sun was coming up. "In May, the days are long, and I could go until after 7:00 p.m." He sailed as much as possible and used the engine or motor-sailed when the waterway narrowed.

I asked Dave if he had any memorable experiences, and he told me that a couple of times trawlers came close to running him down. In both cases, the trawler helmsmen were studying their charts so intently that they didn't see him on their radar screen because his boat was so small. His only resort was to yell as loud as he could. He remembers the horror on their faces as they swerved to avoid him at the last minute.

Included here is an excerpt from his log:

> *Up before 6:00 am and had a difficult time getting the anchor up with the current against the wind. Warmed me up as it was 46 degrees. I bundled up, four layers on top. As a cold front went through it left me with twenty-five knots from the NE and cold air. The wind off St. Catherine's Island and Sapelo Sound* [in Georgia] *gave me a thrashing like I can't remember.*

Dave told me he made a lot of friends on the trip, especially when folks found out how far he was going. "You mean you aren't just going across the bay?" Dave described the experience as "incredibly rewarding." When his wife saw the letters and pictures he received, she said, "Now I understand why you did what you did."

The Modern Rhodes Era

Stuart Scharaga is an entrepreneur from Florida. As a child growing up in Hartford, Connecticut, his knowledge of sailing came from what he read in books. When a friend took him out in a Rhodes on Long Island Sound, he was impressed with how seaworthy the boat was and vowed one day to buy one for himself.

As a successful businessman living in Sarasota, Florida, Scharaga sold commercial real estate and built swimming pools. However, his love affair with the Rhodes 19 remained, and when he had the time, he continued to look for just the right one. He finally found the perfect boat in St. Petersburg and bought it immediately. "I sailed it endlessly," he told me recently.

Meanwhile, the O'Day Corporation had fallen on hard times. In 1980, the Rhodes production facilities were sold to a manufacturer in Michigan and then to a builder with the ephemeral name of Spindrift. When Stuart Scharaga heard Spindrift was planning to retool the boat, he saw a business opportunity. In December 1982, he bought the molds and inventories. Rhodes 19 historian Fred Brehob was impressed with Scharaga's fast action. "Thanks to his dedication and integrity, Rhodes 19 fortunes took a sharp upturn."

Scharaga said, "I put the whole load on a truck and had it driven to Maine, where I planned to open a boatbuilding business." After some negotiations with the Town of Rockport, he set up a business called Stuart Marine. Avant-garde naval architect Jim Taylor was hired to develop production methods and molds that would produce a profitable, sound boat.

Jim Taylor is from Marblehead and is an avid Rhodes 19 racer in his spare time. Jim was on a business trip in Florida when he met Stuart, who had just bought the Rhodes 19 business. Jim says it was simply a matter of being in the right place at the right time when he and Stuart got together.

Stuart Scharaga wanted to improve the way the Rhodes was assembled without fundamentally changing the boat. "We needed to achieve a consistency in the quality. That was a big problem with the O'Day design,"

Stuart told me. Taylor was hired to redesign the way the boat was put together while not changing the class rating.

Jim Taylor told me that the problem with the O'Day design was that it used old-school fiberglass technology: "There were lots of different parts that had to be glassed together, which made it very labor-intensive." Jim came up with a three-piece design: a hull, a deck and an IGU (internal glass unit). His design maintained the boat's aesthetic appearance, and it was easier to produce. The Rhodes 19 Class Rules Committee initially objected, until it tried one out and liked it. It liked it even more after winning a race.

Dave Whittier has owned the Stuart Marine operations in Rockland since the late 1980s. Dave describes himself as "a small businessman in Maine who runs a one-man office. We do whatever it takes to keep the lights on." (Actually, Stuart Marine has a five-man crew.) Dave met Stuart Scharaga in 1982 when he heard Stuart's presentation on building a boat shop at a Rockport zoning board meeting. Afterward, Dave introduced himself and told Stuart he'd enjoy helping him get started. He figured he'd work for Stuart for a few weeks and then go back to his various boat-building jobs.

Stuart's proposal was accepted, and an impressive structure was erected on Route 1. "It was much nicer than what was needed to build boats," Dave Whittier told me. Two years later, Stuart Marine moved its operations to a larger, less expensive site, at an industrial park in Rockland that was just opening up. At the same time, Dave became a full-time employee of Stuart Marine. "Stuart had interviewed me for two years. I guess he decided I might be a good employee." Stuart said he hired David to work for him because he had a great work ethic: "I liked his aggressiveness."

Meanwhile, Stuart Scharaga was beginning to see that for the amount of time and effort he was spending on the boat business, he could have put up four buildings in Florida. He still loved the Rhodes, but he realized that it was easier (for him) to sell commercial estate than to sell a single sailboat. "Sailboat owners are pipe smokers. They take forever to make up their minds," he groused. Stuart added that his wife had come to Maine for couple of months, then said "see you later" and went back to their home in Florida.

The way Dave Whittier saw it was that Stuart discovered that boat building was a tough business. He was also going through his cash very quickly getting the place up and running. What he had thought would be a fun retirement project was turning out to be a grind. Stuart agrees with this

The Stuart Marine crew. Dave Whittier is on the left. *Author's collection.*

assessment: "The boat business is a tough way to make a living. There is a saying, 'If you want to end up with a million in the boat business, start off with two million.' Boy, I found that out to be true."

Stuart Scharaga's business philosophy, however, was important to the initial success of Stuart Marine:

> *You can't invest too much producing your product, or else you end up over-pricing. Fifty years ago George O'Day was the sailboat maven of the country with a complete line of boats. You saw them everywhere you went. But O'Day couldn't make any money because when he tried to cheapen his boat, he took out the classic look...The bigger the boat, the harder it is to get your money back. When you increase your overhead, you have to increase your sales to increase your profits.*

In 1988, Dave Whittier bought Stuart Marine, and Stuart Scharaga returned to Florida to sell real estate. Dave had, however, absorbed the aforementioned business axioms, as well as the importance of setting

Rhodes 19s outside the Stuart Marine boat shop, Rockland, Maine. *Author's collection.*

up multiple profit centers to get his business through hard times. On their website, one sees that, in addition to selling boats, Stuart Marine is a parts business, a brokerage business and a repair and refurbishing business.

Dave Whittier has been building the Rhodes 19 for well over twenty years. When I asked him why the boat has continued to be so appealing, he said, "The Rhodes 19 has a reputation as being very forgiving. You can go out and do stupid things, and the boat will take care of you. It will get you home safely." Stuart Scharaga adds, "The appeal of the Rhodes 19 is that it is both a great starter boat, and that it is also attractive to people coming down from bigger boats."

Fred Brehob calls it "a safe boat with a lot of initial stability. With a wide beam and heavy keel, you really have to work to tip one over." Naval architect Jim Taylor sums it up: "The R-19 story is that the boat has suited so many people so well, in so many different ways, for so long. If I were redesigning it today, I might change some details, but the overall parameters are still pretty darn good."

THE COALING SCHOONER: *WYOMING*

The December 16, 1909 issue of the *New York Times* announced the launching of *Wyoming*, "the largest wooden sailing vessel in the world," built by the Percy and Small yard in Bath, Maine. The ship was named for the state of Wyoming because the governor, Bryant Brooks, was one of the ship's major investors. The launching had taken place on December 14 before a large crowd and went off flawlessly.

The *New York Times* story added, "*Wyoming* is the largest American sailing craft of either wood or steel." *Wyoming* was designed to carry six thousand tons of coal, more tonnage than the entire fleet of vessels that Bath shipbuilder Jonathan Davis owned at the start of the nineteenth century. *Wyoming*'s hull was 350 feet long; by adding the spanker boom, the ship was 450 feet long. From keel to the top of her masts, the great schooner was taller than a seventeen-story building, and she drew 27 feet when fully loaded. The big ship represented a huge investment ($190,000) for the Percy and Small yard. Given the uncertain future of the coastal coaling business, she was also the cause of considerable anxiety.

In the late nineteenth and early twentieth centuries, schooners were the oceangoing workhorses for New England, including Maine. Lumber for shipbuilding and coal for industrial use were in demand as the population of the area boomed after the Civil War. There was no set number of masts for a schooner, but the ships were easy to work. Roger Duncan, in *Coastal Maine*, writes, "The schooner rig with its fore and aft sails required only a small crew. It was especially well adapted for carrying bulk cargoes along the Atlantic coast." The two-masted schooner, in particular, became known as "the errand boy" of the coast. As demands for moving cargoes increased, two-masted schooners became larger, and by the mid nineteenth-century, shipbuilders were adding a third mast.

As the century neared its end, the success of these ships led to the building of four-, five- and six-masters. The value of the "great schooners," as they were called, lay in the efficiency of their rig. Lincoln Paine tells us, "Large schooners achieved significant economies of scale with an average of two crewmen per mast. The remainder of the crew consisted of a mate, an engineer or cook, and the captain." *Wyoming*, for example, carried a crew of thirteen. John Hutchinson summed them up: "The great American schooner was the most weatherly and economical sailing vessel in the world."

From 1879 to 1921, over five hundred great schooners were built on the Atlantic and Gulf coasts. Of these, three-quarters were built in Maine.

Launching of *Wyoming*. *Courtesy of Mariners Museum, Newport News, Virginia.*

Bath, in particular, was a major center for schooner building. With a dozen shipyards, the town launched over thirteen hundred schooners of all sizes during the nineteenth century. In 1882, for example, Bath employed four thousand men in its twelve yards and launched more wooden vessels than any other port in the world. Narrowing the focus further, Percy and Small, one of the largest yards in Bath, specialized in five- and six-masters. From 1894 to 1920, it launched forty-two big schooners.

Although most of Maine's great schooners were built in Bath, four-, five- and six-masters were also launched in the deepwater ports of Thomaston, Rockland, Rockport and Camden.

Roger Duncan informs us that six-masted schooners approached the practical limit for wooden vessels. *Wyoming* was built of six-inch-thick yellow pine planks, supported by ninety iron cross bracings on each side. Despite the support, and largely because of its extreme length, *Wyoming*'s hull tended to flex in heavy seas. This caused the planks to bend, thereby allowing seawater into the hold. The result was that *Wyoming*'s pumps were constantly in use in an effort to keep the hold free of water.

Wyoming's accommodations were "comfortable and cozy," wrote John Fuller in the *Portland Telegram*. Officers' living quarters were "enhanced by steam heat, electric lights, telephone communications and plush furnishings." Navigationally, *Wyoming* was equipped with a modern sounding device, steering controls and high-pressure gauges. Fuller added, "The schooner was endowed with about everything except internal propulsion," which was not considered necessary for vessels carrying "low-priority" cargoes of coal, lumber and gravel.

Wyoming carried an auxiliary steam engine for raising and lowering sails, pumping out water and working the Hyde anchor windlass. The result was a highly efficient ship that, at times, was worked by as few as eleven men. All in all, she was considered the ultimate schooner.

Ralph Snow, in *A Shipyard in Maine*, informs us that under the "meticulous and experienced command of Captain Angus McLeod," *Wyoming* "was a good sailer, handled easily and responsively to her helm." To quote Captain McLeod, "She is as handy a yacht and the best working vessel I have put my foot into." Ralph Snow continued, "*Wyoming* avoided all the perils that appeared to lie in wait for her fellow coal schooners. Groundings were unheard of, as were lost anchors and limping into port with sails blown out."

Wyoming underway. *Collections of Maine Maritime Museum, courtesy of Maine Memory Network.*

With her full spread of thirteen thousand square yards of canvas in a good breeze, *Wyoming* was faster than most freighters of the day. At one point during the First World War, she crossed from New York to France in eighteen days, reportedly eluding more than one German U-boat during the course of the voyage.

Captain Angus McLeod skippered *Wyoming* until Percy and Small sold the ship to the France & Canada Steamship Company in 1917 for $400,000. With the appearance of coal barges and steam colliers in the coal shipping business, Percy and Small had to face economic reality and began to sell off its fleet of aging, and expensive to maintain, five- and six-masters. The competition from tug-propelled barges and colliers was formidable. With smaller crews, they were more economical to run and could follow a fixed schedule. In addition, they could carry a cargo of seven thousand tons.

At the end of the First World War, *Wyoming*'s new owners leased her to carry coal to European ports. One of the ship's skippers during these years was Captain Charles Bullock. In a letter to his wife, he gives us an idea of what an Atlantic winter crossing was like:

> *January 1919: My Darling, Today is nasty. We got in a north-west gale and lost our mainsail…The wireless is trying to get a* [passing] *steamer to relay a message for us to New York, but we are too far away to send one…This morning a steamer signaled us and the mate did not think to call me when we could have got our position…Bermuda: I am now in the St. George Hotel, just as nice as could be. I am glad to be here as it was very bad at sea. The Lord is good to me this time…We have got to have new sails which means sending to New York for them and that will mean about two weeks here…with lots of love to all from your husband.*

We don't know much about Charles Bullock, although from the tone of the letters to his wife and children, he appears to have been a devoted husband and father. As captain of *Wyoming*, however, his challenges were not always nautical in nature. A story in a Buenos Aires newspaper entitled "Too Free with the Knife" reports:

> *Captain Bullock, of the six-master* Wyoming, *was stabbed by one of his crew, a Negro. From the facts reported, it appears that the cook had been reprimanded by the captain because he left the vessel while she was in port and had not provided the meals. Members of the crew pulled the Negro off the officer whose face was criss-crossed with cuts. The port*

Wyoming waiting to be loaded with coal, circa 1917. *Photo from* New York Times *archives.*

> *police were summoned and the knife-wielding gentleman was taken off to prison. Captain Bullock was in a dangerous condition for some days, but fortunately is now on the way to recovery.*

In 1923, Charles Barth, a young Norwegian, signed on as a member of *Wyoming*'s crew for thirty dollars per month. He recalled his surprise that eight sailors, two mates, a cook, an engineer and a captain were all it took to manage the huge vessel. Barth also recalled watches of six hours on and six hours off and spending a lot of time wrestling with the large, and very difficult, wooden steering wheel. He also remembered being perpetually hungry.

After several trips, Barth came down with typhoid fever and was put ashore at the Chelsea Marine Hospital in Boston. By the time he recovered, *Wyoming* had departed for her final, fatal voyage

The great schooner met her end in March 1924 in a violent storm that pounded the North Atlantic coast. *Wyoming* and another schooner, *Cressy*, anchored in a stretch of water that separated Nantucket from Cape Cod, known as Pollock Rip, hoping to ride out the gale. After two days of pounding, *Cressy* weighed anchor and headed out to sea. *Wyoming*, however, remained at anchor.

The next day, a "profusion of wreckage" began coming ashore on Nantucket Island, including a nameplate labeled *Wyoming*. Had the great ship simply come apart when the powerful winds opened gaps in her planking? Did she collide with another vessel also reported missing in the storm? What happened to the crew? Two lifeboats were launched, yet none of the thirteen crewmen was ever seen again.

Wyoming's loss hastened the demise of the great schooners. A few of the huge ships carried on, though no new ones were built after the First World War. Schooners were already losing the competition with coal barges and steam colliers. Each would soon surrender to the economic and technological advantages of the internal combustion engine as a means of water and land transport.

The Arctic Schooner: *Bowdoin*

Three years before the sinking of the schooner *Wyoming* in 1924, the Arctic schooner *Bowdoin* slid down the ways of the Hodgdon boat yard in East Boothbay, Maine. It was April 21, 1921, to be precise. And so, as the tale of one schooner ends, the story of a very different schooner begins.

The story of *Bowdoin* is the story of the people who have sailed her: Admiral Donald MacMillan and his wife, Miriam; Jim Sharp; Andy Chase; and other skippers over the years. The list concludes with Eric Jergenson, *Bowdoin*'s present small craft master at the Maine Maritime Academy. What is most impressive about *Bowdoin* is the incredible loyalty and devotion the ship has inspired in almost everyone who has come in contact with her.

The MacMillan Years

Bowdoin was built in 1921 by Donald "Dan" MacMillan, who took the ship on more than two dozen voyages of Arctic exploration. MacMillan, who was born in Provincetown in 1870, was the son of an intrepid skipper who perished at sea in 1883. When his mother died a few years later, Dan and his four siblings were orphaned.

Undaunted, MacMillan worked his way through Bowdoin College, graduating in 1898. Following a ten-year stint as a teacher, he joined Robert

E. Peary's expedition to the North Pole in 1908 and began his passionate lifelong career as an Arctic explorer.

One winter, while awaiting a relief ship in Greenland, MacMillan began to think about the characteristics of a good Arctic vessel. Following a stint in the navy during World War I, MacMillan began raising the money to build his own Arctic schooner. *Bowdoin*, named after his alma mater, was completed in 1921, and MacMillan's career as an Arctic explorer was launched.

Designed expressly for working in the Arctic, the two-masted *Bowdoin* is eighty-eight feet long, twenty-one feet wide and has a ten-foot draft, which allows her to work close to shore. Present captain Eric Jergenson describes her as "a Bald Headed Knockabout Schooner, with no topsails or bowsprit." She was double planked for added strength and has an auxiliary engine, since the Arctic is known for having either no wind or too much. The hull is rounded, which allowed the ship to rise out of the water when caught between ice floes. The bilge was filled with cast concrete for added strength. An 1,800-pound steel nose plate is bolted to the hull to help *Bowdoin* in its passage through heavy ice.

Bowdoin's maiden voyage (1921–22) was sponsored by the Carnegie Institute and proved to MacMillan that he had designed an ideal ship for Arctic exploration. With poor or nonexistent charts, the schooner survived encounters with ledges and icebergs and crossed the Arctic Circle on August 23, 1921. *Bowdoin* spent a long winter at Baffin Island, where she was frozen in for 274 days. Numerous scientific and meteorological observations were made from her position, including a low of minus fifty degrees on February 10, 1922.

For the next fifteen years, MacMillan made repeated voyages to Labrador, Greenland and beyond, surveying, mapping and collecting hundreds of botanical, zoological and geological specimens. In addition, he studied Inuit (Eskimo) life and language and established an Inuit school in Nain, Labrador. In all, MacMillan would sail *Bowdoin* over 250,000 miles on twenty-six northern voyages from 1921 to 1954.

During the 1920s, *Bowdoin* carried a diverse crew of experienced sailors, scientists and adventurers on her Arctic voyages. In the 1930s, in order to help finance his trips, MacMillan began to take paying college students, often from Bowdoin College. In her book *The Arctic Schooner* Bowdoin, Virginia Thorndike writes, "Each boy signed a contract acknowledging the complete authority of MacMillan ashore and afloat until the voyage shall have ended." In the process, Mac turned each young man into a capable sailor.

Maine's governor, Owen Brewster, with Commander MacMillan as *Bowdoin* sets forth on another Arctic expedition. Wiscasset, June 19, 1926. *Collections of Maine Historical Society/ Maine Today Media, courtesy of Maine Memory Network.*

In 1935, sixty-year-old Dan MacMillan married twenty-nine-year-old Miriam Look, the daughter of his closest friend. MacMillan had never had a woman on board, and it was three years before Miriam was able to persuade her husband that she was up to the rigors of an Arctic voyage.

The original helm and binnacle is still in use on the *Bowdoin*. *Author's collection.*

"Lady Mac," as she was called, proved to be a "good scout," according to a crew member, and she accompanied her husband on *Bowdoin*'s next nine voyages. What was her secret? "After eight trips with him I've learned not to ask questions. I also try to answer questions of others before Mac hears them."

In 1941, *Bowdoin* was sold to the navy for the duration of World War II. In May 1942, Lieutenant Stuart Hotchkiss took command of the schooner, and for the next year and a half the vessel was assigned to the South Greenland Patrol. This involved setting up air bases on Greenland, performing hydrographic surveys, mapping shorelines and taking soundings, all the while avoiding German U-boats. At the end of 1943, Lieutenant Hotchkiss was transferred to destroyer escort duty, and *Bowdoin* was sailed to Boston, where she was decommissioned for the rest of the war.

In a chapter entitled "Back to Mac," Virginia Thorndike writes that *Bowdoin* had become "a derelict" by the end of the war. In 1945, seventy-year-old Admiral Donald MacMillan bought his beloved ship back from the navy for $4,000 and began to restore her. A year later, *Bowdoin* was ready for a shakedown cruise to Labrador. Virginia Thorndike tells us, "Instead of his usual student crew MacMillan had a mixed group, with only two paid members. The rest were 'vacationers'; people taking breaks from their normal lives." The thirty-five-day trip was a success, and *Bowdoin* was ready to go back to work.

For the next nine years, Admiral MacMillan continued to take *Bowdoin* on scientific expeditions to Arctic waters. He made his last trip in 1954 at the age of eighty. Over the years, the MacMillan expeditions made extensive contributions to our knowledge of Arctic regions. In addition to scientific studies too numerous to mention, Mac demonstrated that airplanes could be used above the Arctic Circle and that short-wave radio provided effective communication with the rest of the world.

The Post MacMillan Years

By the mid-1950s, MacMillan had reached the age where he was no longer able to take *Bowdoin* on Arctic expeditions. There were those who felt the aging schooner should be left to rot, but the old admiral was anxious that the vessel be properly cared for. Accordingly, he and Miriam spent several years looking for a suitable port. Finally, in 1959, a home was found at the Mystic Seaport Museum in Connecticut, which had recently opened an Arctic exhibition.

There was just one problem: the museum had no money to purchase the vessel. Following a major fundraising effort by the MacMillans, their friends and Bowdoin College alumni, enough money was raised to allow the museum to purchase the old schooner. MacMillan, now eighty-five, sailed *Bowdoin* to Mystic, Connecticut, and on June 27, 1959, she was officially welcomed to the Seaport Museum by a large crowd. Virginia Thorndike wrote, "It seemed like the perfect home for the schooner."

Less than ten years later, a new director with different priorities decided to dismantle the Arctic exhibition. Maintenance on *Bowdoin* had been neglected, with the result that the schooner lay rotting at her berth. The vessel was permanently closed and sheathed in plastic, which in the summer heat accelerated the decay of her timbers.

Bowdoin's plight did not go unnoticed by Admiral MacMillan, who was appalled by her condition. In 1968, MacMillan went public with his concerns and announced that he would give *Bowdoin* to anyone who could care for her. At this point, Captain Jim Sharp stepped forward and agreed to tow the vessel back to his dock in Camden, Maine.

At about the same time, a number of people who had sailed on the historic schooner formed the nonprofit Schooner Bowdoin Association. Long-range plans were vague, but the idea was to save *Bowdoin* from further neglect. After some negotiations, Jim Sharp leased *Bowdoin* for a dollar a year from the Schooner Bowdoin Association, with the understanding that he would begin repairs on the historic vessel.

Over the next several years, *Bowdoin* was restored by Jim Sharp and a dedicated volunteer, John Nugent. A new foremast was stepped, planks were replaced, paint was stripped and years worth of barnacles were scraped off the hull. In all, Sharp estimates that he spent $30,000 (in 1960s dollars) of his own money getting *Bowdoin* ready for sea.

One of Sharp's first objectives was to sail *Bowdoin* to Provincetown to show her to MacMillan while the old admiral was still alive. When they arrived in the fall of 1969, the MacMillans were in tears, Jim said. "Before long, half the town was on the boat. We had to stop people

In 1968, Captain Jim Sharp towed *Bowdoin* to his dock in Camden for a major overhaul. *Courtesy of Jim Sharp.*

from coming on board to keep it from sinking." Less than a year later, MacMillan died.

In 1975, relations between Jim Sharp and the Schooner Bowdoin Association broke down. Captain Jim and his family had a wonderful seven years of cruising and restoring the old vessel, but Sharp couldn't afford to keep *Bowdoin* when the association insisted on ending the lease. "If she'd been mine, I'd be sailing her to this day," he told me recently.

The Schooner *Bowdoin* Inter-Island Expeditions was formed in 1976, and for the next few years the ship was used for educational programs. The schooner took students on trips from Penobscot Bay to Canada, introducing them to marine biology, botany, ornithology and underwater archaeology. At the end of the 1979 season, however, the directors of the program were told that *Bowdoin* would need major repairs to meet new Coast Guard regulations.

There were those who thought *Bowdoin* should be scrapped, but Miriam MacMillan and the Schooner Bowdoin Association held out for a rebuild. As a result, from 1980 to 1984, a major restoration of the vessel took place, mostly on the grounds of the Maine Maritime Museum in Bath, the site of the old Percy and Small boatyard. At one point, the money ran out, but such was *Bowdoin*'s magnetism that skilled volunteers kept showing up on weekends to finish the job. When completed, the restored schooner was remarkably similar to MacMillan's original design.

For a number of years, the Bowdoin Association had been looking for a group to assume responsibility for the schooner. In the meantime, the old vessel was being used hard. Virginia Thorndike notes, "During a period of eighteen months more than two thousand people participated in a variety of programs on *Bowdoin* ranging from Maine to Chesapeake Bay."

In 1988, *Bowdoin* finally found what appears to be a permanent home. Ken Curtis was president of the Maine Maritime Academy in Castine when he heard that *Bowdoin* was available. He quickly realized the potential the vessel offered both as a student training ship and for public relations. Accordingly, Maine Maritime sold one of its little-used sailing yachts and put the proceeds into *Bowdoin*'s purchase.

Bowdoin long ago achieved celebrity status. Everywhere she goes, the ship is greeted by admiring crowds. On August 4, 1988, the schooner was designated Maine's official state vessel by the governor, and in 1989 she was made a National Historic Landmark.

Bowdoin underway in Penobscot Bay. *Courtesy of Maine Maritime Academy.*

Bowdoin's captain, Eric Jergenson, at the Maine Maritime Academy. *Author's collection.*

As a training ship, *Bowdoin* is a hands-on classroom for specific academic courses. Captain Eric Jergenson also takes students on longer trips to Labrador and occasionally to Greenland. As an ambassador for Maine Maritime, *Bowdoin* is made available to community organizations whenever possible. Wouldn't Admiral MacMillan be pleased with the many ways his historic old schooner continues to work her magic?

IV

Storms and Shipwrecks

Hurricanes in Maine

Early Gales and Tempests

Hurricanes don't begin in Maine. Starting in southern climes, they usually lose their intensity the farther north they get, and many end up heading out to sea. There have been, however, some notable exceptions.

It was not until 1898, when President William McKinley ordered the Weather Bureau to establish a hurricane-warning network, that the United States began to keep accurate records of these storms. And it was not until 1963 that David Ludlum, a weather historian, published a book entitled *Early American Hurricanes, 1492–1870,* based on his research of old letters and diaries. What follows is a brief summary of some of the particularly bad storms that hit Maine before the late nineteenth century.

The Great Colonial Hurricane occurred in August 1635. The hurricane followed a track similar to the 1944 Great Atlantic Hurricane and Hurricane Edna in 1954, both of which devastated parts of Maine. Governor John Winthrop of Massachusetts wrote, "There was such violence that blew down many hundreds of trees…overthrew some houses and drove ships from their anchor."

Just off Pemaquid Point, the full force of the gale caught *Angel Gabriel,* carrying one hundred settlers from England. The ship was wrecked on the rocks, although fortunately only five passengers were lost. As of 2003, meteorologists consider the 1635 tempest to be one of the five "most intense hurricanes" to have made landfall in New England.

The 1635 tempest was followed by a storm in September 163 that John Jocelyn of Scarborough, Maine, then part of Massachusetts, described as follows:

> *About 4-o'clock in the afternoon, a fearful storm of wind began to rage, called a hurricane...The greatest mischief it did us, was the wracking of our shallops* [small boats], *and the blowing down of many trees, in some places a mile together.*

Beginning in 1769, the Reverend Thomas Smith of Portland described several storms that affected Maine. Smith's letters describe the 1769 event as "a dreadful storm." Rhode Island sustained the most serious damage, but Maine experienced "heavy rain and wind causing tree limbs to break." Then, in 1770, an "exceedingly great storm passed through the area with a low pressure of 28.96." And in 1788, Smith tells us, another storm, known as the Western New England Hurricane, created a wide path of destruction. "It blew down houses and barns, trees, corn and everything in its way. Such a hurricane as was never the like in these parts of the world."

At the conclusion of the War of 1812, the Great September Gale of 1815 roared through New England. It was the worst storm to hit the coast since 1635, and it was not until the 1938 hurricane that a storm of such intensity would create so much havoc. Edward R. Snow wrote in *Storms and Shipwrecks of New England*, "It was the most severe storm of the nineteenth century." When the storm hit Maine, Snow tells us, "in Wells a man was killed by a tree felled by the wind and much damage was done to buildings."

The September Gale in 1869 was the first of two storms to hit Maine that fall. In Portland, newspapers reported that it was the most severe storm the city had ever known. The steeple on the cathedral was blown down, causing $6,000 in damage. In addition, chimneys, telegraph wires and tree branches littered the streets. Train service was delayed for quite some time because of obstructions on the tracks. And in nearby Bath, damages were estimated at between $25,000 and $50,000, in 1869 dollars.

Along the coast, the September Gale sank thirty ships, and in Casco Bay, *Helen Eliza*, a schooner out of Rockport, Massachusetts, was wrecked off Peaks Island with a loss of eleven of her crew.

In October of that same year, a second storm hit Maine, known as Saxby's Gale. Wayne Cotterly described the origins of the name in *Hurricanes & Tropical Storms and Their Impact on Maine*:

> *The storm's name* [Saxby's Gale] *came from a British Naval Lieutenant by the name of S.M. Saxby who predicted in November 1868 that a storm would occur with unusual violence and high tides on October 5, 1869. This prediction was made due to the proximity of the moon to the earth's equator. Saxby had previous success with these predictions, and was taken somewhat seriously.*

Washington County took the brunt of Saxby's Gale, but it was felt all along the coast. A strong storm surge affected many of the rivers and tributaries up to fifty-five miles inland. At Fredericton, New Brunswick, the tide was reported to have risen three feet. Up to six inches of rain fell in some areas, and the Androscoggin River rose to its highest stage since the floods of 1832.

A man traveling from Eastport to Calais reported that the gale blew down or seriously damaged ninety houses. Another man wrote, "There never was such a gale hereabouts as this since the country was settled...The tide rose beyond all precedent."

The 1938 and 1944 Hurricanes

Before we examine some of the twentieth-century hurricanes that have affected Maine, it is important to review the categories of hurricanes that were developed by the National Hurricane Center in 1972. Known as the Saffir-Simpson scale, the categories are based on the intensity of wind and were established by civil engineer Herbert Saffir and meteorologist Bob Simpson, director of the National Hurricane Center:

> *Category One: Sustained winds 74–95 mph, dangerous winds that will produce some damage.*
> *Category Two: Sustained winds 96–110 mph, extremely dangerous winds that will cause extensive damage.*
> *Category Three: Sustained winds 111–130 mph, devastating damage will occur.*
> *Category Four: Sustained winds 131–155 mph, catastrophic damage will occur.*
> *Category Five: Sustained winds 155+ mph, catastrophic damage will occur.*

Over the past seventy-five years, Maine has had seven hurricanes that have caused major damage. Of these, the 1938 category five storm that hit the Atlantic coast was by far the worst, leaving sixteen thousand people homeless and causing six hundred deaths and $3,593,853 worth of destruction ($4.7 billion in 2010 dollars).

What is interesting about the New England Hurricane of 1938 was that it had been a generation since a storm of major significance had hit the northeast Atlantic coast. People had begun to associate hurricanes with warmer climes like Florida and the Gulf Coast when a strong tropical depression developed off the coast of Africa late in the summer of 1938.

The 1938 hurricane remains the most destructive storm in New England's recorded history. It took twelve days to cross the Atlantic Ocean and grow into a category five hurricane. It passed over Long Island and hit Providence, Rhode Island, on the morning of September 21 with wind gusts of 160 miles per hour. The Weather Bureau had followed the storm, but believing that it was headed out to sea, it had issued no warnings, except for small craft advisories.

Rhode Island and the city of Providence were inundated by a storm surge fifteen feet above normal, made worse by the fact that it was high tide when the gale hit. Bridges and railroads were washed out, and dams were breached. Coastal New England was devastated.

In Maine, the damage was particularly serious in the Lewiston-Auburn area, where the storm headed inland. Five people were injured, though miraculously no deaths occurred. The *Lewiston Daily Sun* called it "the worst wind storm in twin cities [Lewiston-Auburn] history." Power went out at 10:30 p.m. on September 22, and dispatchers were kept busy throughout the night answering emergency calls. At one point, downed trees and debris leading out of the area blocked virtually every road. The thousands of trees that were blown down constituted a fire hazard in Maine's forests for many years to come.

The Great Atlantic Hurricane of 1944 developed in the Leeward Islands of the Caribbean early in September 1944. It moved slowly northwest, intensifying into a category four hurricane by the time it reached Cape Hatteras. The hurricane skirted the East Coast and made landfall on Long Island, where it continued to move northeast, passing over Boston and into the Gulf of Maine. From there, it crossed southeastern New Brunswick, eventually reaching Newfoundland.

As the storm passed Maine on September 14, it was accompanied by torrents of rain. In Portland, wind gusts reached sixty miles per hour.

Rain flooded streets, trees were uprooted, falling limbs damaged cars and buildings and there were widespread power outages. In Androscoggin County, 40 percent of the apple crop was destroyed by high winds, and at Bates College, 4.34 inches of rain were recorded.

The hurricane took two lives in Maine, including a ten-year-old boy who was electrocuted by a downed power line. The storm claimed the lives of 390 people, most of whom were military casualties. Remember, the United States was in the middle of World War II. Two Coast Guard cutters were sunk off Cape Hatteras with a loss of 48 men, a minesweeper went down with 33 on board and the destroyer USS *Warrington* sank off Vero Beach, Florida, with a loss of 248 sailors.

Although forty-six deaths occurred on land, the figures could have been much higher. One of the major differences from the 1938 hurricane was that civil defense officials were able to follow the storm's track. With advance warning, they were able to evacuate thousands of people from low-lying areas.

The Great Atlantic Hurricane of 1944 resulted in over $100 million in damages, $1.2 billion in 2010 dollars. The fifty- to seventy-foot waves crashing against the coastline left debris scattered along miles of beaches and destroyed thousands of shore properties and small boats.

As a postscript to this account, I would like to add a personal note. As the hurricane crossed Long Island Sound and moved toward Rhode Island, the Coast Guard went house-to-house issuing warnings for coastal residents to evacuate. My wife, who was then four, and her one-year-old brother were spending the summer with their grandmother in Narragansett, Rhode Island. That intrepid lady had taken her grandchildren down to the beach to look at the rough seas. As a result, they missed the Coast Guard warning and returned to their house without realizing the severity of the storm or the need to evacuate.

According to my wife, when the storm struck in the afternoon, the sky turned yellow. Later in the evening, she remembers her grandmother standing on a stepladder nailing boards across the windows that faced the sea to keep water from coming in. Needless to say, it was a tumultuous night. The next morning, the lawn was covered with boulders pushed up from the nearby shoreline by the raging seas.

The 1954 Hurricane Season: Carol, Edna and Hazel

The 1954 hurricane season was a very active one for the northeastern United States. There were a total of twelve named storms, eight of which were hurricanes. Three of these—Hurricane Carol, Hurricane Edna and Hurricane Hazel—were major storms. The combined storms of 1954 caused 1,069 deaths and cost $751.6 million ($6.08 billion in 2010 dollars).

Hurricanes Carol (category two) and Edna (category three) followed nearly identical paths, brushing the North Carolina coast before hitting New England. Maine had been spared from serious destruction since the Great Atlantic Hurricane in 1944, but the destruction caused by Carol and Edna totaled $500 million in damages, $25 million of which were in Maine. What made it even worse was that the two storms were only ten days apart.

Hurricane Carol began as a tropical depression in the Bahamas in late August 1954. The storm drifted erratically as it moved up the coast. Carol made landfall as a category three hurricane on eastern Long Island with winds of 120 miles per hour. As it crossed southern New England, Rhode Island and Cape Cod, especially, were badly battered. (I should point out that this was the storm that devastated the SMYRA racing class on Edgartown, Martha's Vineyard that Peter Plumb was referring to in the Rhodes 19 story.) Thousands of buildings were swept away in a ten- to fifteen-foot storm surge, and in Boston, the spire of the historic Old North Church was blown down.

Carol maintained its intensity as it moved into Maine. Winds were measured at eighty miles per hour at the Augusta airport and seventy miles per hour in Portland. In the Twin Cities area of Lewiston-Auburn, the damage was severe. Apple orchards were flattened, the corn crop was destroyed, uprooted trees crushed cars, telephone poles were knocked down, houses were damaged and there were widespread power outages. At the time, Hurricane Carol was the worst natural disaster in Maine's history. Unfortunately for the state, this distinction would be short-lived.

Sixty-five people died from Hurricane Carol in New England, three of whom were in Maine. As the cleanup began, people nervously watched Hurricane Dolly moving up the coast, a storm that fortunately moved out to sea. Less noticed, and as yet unnamed, was another storm. Edna would strike the East Coast ten days later.

Edna formed off the coast of South America in early September. It skirted the Bahamas and followed a similar track to the one Carol had taken as it passed to the west of Florida and Georgia. Edna passed to the west of

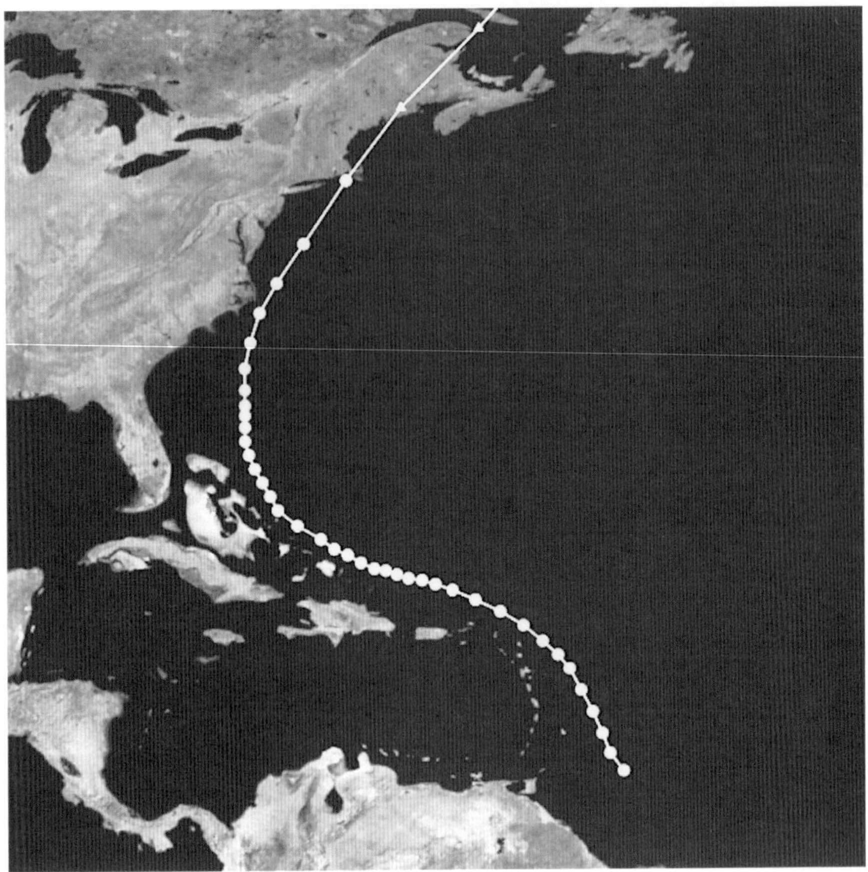

Map showing the course of Hurricane Edna. *Wikipedia.*

Cape Hatteras. With winds of 125 miles per hour, it was a category three hurricane, even more powerful than Hurricane Carol.

As the eye of the storm passed between Nantucket and Martha's Vineyard on November 11, it split into two sections, an unusual meteorological event. Low-pressure readings at Truro (28.29) and Nantucket (28.18) supported the presence of two eyes, over sixty miles apart. On Martha's Vineyard, wind gusts were reported at 120 miles per hour.

Hurricane Edna then crossed Cape Cod and swept into the Gulf of Maine. Edna was considered a "wet" storm, with the result that Maine got drenched. Parts of the state received seven to eight inches of rain, and winds were in the sixty- to seventy-five-miles-per-hour range in the Portland area. The Kennebec and other rivers flooded, and raging waters washed out bridges and roads, uprooted trees and knocked down power lines throughout southern and central Maine.

Eight people died due to drowning in the state. In Unity, Maine, a family of ten were trapped on the top of their car, cut off by furious floodwaters. A human chain managed to save nine of them. In all, Hurricane Edna claimed twenty-nine lives in the United States. Damages were estimated at $15 million in Maine and $40 million throughout New England.

I had left Maine to return to Massachusetts for my senior year of high school by the time Hurricane Edna began to advance up the Atlantic coast in September 1954. However, my friend and neighbor on Vinalhaven, seventeen-year-old John Washburn, was still on the island. When we were

Effects of Hurricane Edna on inland Maine, as seen in the flooding of the Royal River. *Collections of North Yarmouth Historical Society, courtesy of Maine Memory Network.*

discussing the event recently, he remembered that Edna's track passed over Penobscot Bay. Fifty-five years later, John still has powerful memories of the storm that hit Vinalhaven on September 15, 1954:

> *I had gone to bed and was sleeping in an upstairs bedroom that had a small window that looks out to the east. I was not quite asleep because of enormous noise; the howling of the wind, the thrashing of trees and crashing of waves. Suddenly there was a complete silence and I remember wondering how the hurricane could be over so quickly, and why, suddenly, there was moonlight. I went to the window and saw a large patch of clear sky with clouds circling rapidly around. In the middle was the moon, which illuminated the clouds. Otherwise, except for a few stars, the sky was jet black.*
>
> *What was extraordinary was that I was actually seeing the eye of the storm passing at night. It was just incredible luck that I got to the window when I did. The whole scene was very beautiful and it had a very strong aesthetic impact on me. The storm was moving fast and within a couple of minutes the eye had passed over and the hurricane resumed.*
>
> *The next morning trees were down all over the place. You could see where the most intense track of the hurricane had been as it crossed the island.*

Although it didn't hit Maine, the effects of Hurricane Hazel need a brief mention. Coming on the heels of Carol and Edna, it was the deadliest and costliest storm of the 1954 season. Hurricane Hazel was a category four hurricane that first devastated the Caribbean Sea in early October 1954. It came ashore in North Carolina on October 15, with winds estimated at 140 miles per hour, and created a path of destruction through the eastern United States as far north as Canada. When Hazel arrived in Ontario, rivers and streams overflowed their banks, causing severe flooding. Hazel was particularly destructive in the Toronto area as a result of a lack of experience in dealing with tropical storms and the storm's unexpected retention of power. Because of the devastation it caused, the name Hazel was retired.

John Washburn as a teenager. *Author's collection.*

The name Carol was reused in the 1965 season but was retroactively retired when the modern system of alternating male and female names was introduced in 1979. The name Edna was used again in 1968 and then retired.

Bob and Irene

Hurricane Bob made landfall in New England in late August 1991 after missing most of the southeastern Atlantic coastline. The category two storm caused $26 million in damages in Maine and $1.5 billion overall. It claimed eighteen lives, including three people in Maine. In Cumberland, a sixty-year-old man was swept away by floodwaters trying to escape from his disabled truck, and in South Portland, a man was electrocuted while connecting a water pump in a flooded basement.

Strong winds and 7.8 inches of rain battered the Portland area. At Blue Hill, peak winds were clocked at ninety-three miles per hour, and winds of sixty-nine miles per hour were recorded at Mt. Desert Island and at Matinicus Rock.

Wind gusts from Bob ripped roofs off homes, boats from moorings and utility poles out of the ground, leaving hundreds of thousands of people without power. In Topsham, Mike's Ice Cream store was destroyed by a power surge when it caught fire.

I was on Vinalhaven as Hurricane Bob moved up the coast, and I recall the anxiety we all felt as the storm headed on what seemed like a collision course with Penobscot Bay. We secured our boats and nervously listened to weather reports as the storm moved closer. Then, at what seemed like the last minute, Bob headed inland, slightly south of Augusta.

Although Maine escaped relatively unscathed, in 2011 Hurricane Irene devastated parts of New England, especially Vermont, causing massive power outages, flooding and property destruction.

Once again, we breathed a sigh of relief—although, as anyone who lives on the Gulf of Mexico or the Atlantic seaboard knows, the threat of hurricanes is always present.

The Burning of *Royal Tar*

When I was a lad my Granddad told
Tales of pirates and shiploads of gold,
Buried treasures and missing ships,
Fisherman's spirits whose whitened lips
Seemed repeating the words beware, beware!
Of the sunken reefs and the dangers there.

But the story he told that gripped us tight
Was told on a stormy summer's night,
'Midst the shrieking wind and thunder's jar,
He spun the yarn of the "Royal Tar,"
An ill-fated steamship as you have read,
Stormbound and anchored off Coombs' Head.
—*Anon.*

In 1836, the owners of the Eastern Steamboat Line proudly announced their latest acquisition: "the new and superior steamer *Royal Tar*." The ship was named after King William IV of England, who was known to be an avid sailor. Weekly service between St. John, New Brunswick, and Portland, Maine, began in the spring of 1836, when *Royal Tar* joined the Eastern Line's four other steamers on a network of routes that ran along the New England coast from Canada to Boston. *Royal Tar* ran for one season on the Eastern Line before it was destroyed by fire off the eastern shore of Vinalhaven Island in Penobscot Bay.

The four-hundred-ton *Royal Tar* was 160 feet long and had a 24-foot beam. The wooden side-wheeler was built in Carleton, New Brunswick, across the river from St. John, at a cost of $50,000. *Royal Tar* drew 12 feet and was considered both a schooner and a steamship, with the advantages, and disadvantages, of each.

In the 1830s, steamboat engines were often unreliable, and sails provided a backup in case an engine broke down or the ship ran out of fuel. *Royal Tar* was driven by two paddle wheels, considered a more efficient means of propulsion at the time than propellers, which tended to vibrate, causing leaks in the shaft area. Sails were used for longer voyages, since early steam vessels couldn't carry enough fuel, usually wood, for a voyage of more than a day or two.

Reversing falls, St. John River, New Brunswick. *Royal Tar* was launched on the right side of the river where it starts to curve toward the harbor. *Courtesy of Mark Warner.*

It is important to note that in 1836 there were no restrictions regarding the number of passengers or lifeboats a passenger vessel needed to carry. Nor did life jackets exist, in the formal sense, until cork ones appeared in the 1850s. On her fateful trip, *Royal Tar* carried approximately ninety-three passengers and crew, in addition to a circus of two dozen wild animals and birds. There were thirty-four cabins on deck for the well-to-do; steerage passengers slept on deck or bunked below.

Mark Warner, in his recent book *The Tragedy of the* Royal Tar, informs us that in the 1830s, European immigrants were looking for a faster way to get to the United States from Canada. The long, uncomfortable land trip by stagecoach might take a week or more. On the other hand, a steamer could make the trip from New Brunswick to Portland, Maine, in a day and a half. From there, Warner tells us, passengers could take a ship to Boston or New York. On its last voyage, many of *Royal Tar*'s seventy-two passengers were Irish immigrants, among whom were sixteen women and twelve children.

In the summer of 1836, *Royal Tar*'s schedule ran between St. John, Canada, and Portland, with a regular stop at the border town of Eastport, the easternmost tip of the continental United States. Heading south from Eastport, *Royal Tar* would pass by Jonesport and Mt. Desert and, depending on the weather, run outside of Matinicus Rock, Monhegan Island and Damariscove Island before entering Casco Bay and Portland Harbor. In rough weather, *Royal Tar* took the longer, more sheltered route across Penobscot Bay and between the North and South Fox Islands (today North Haven and Vinalhaven) rather than go around Matinicus Rock.

Thomas Reed, who was considered to be a skillful mariner, skippered *Royal Tar*. One of his passengers wrote, "His gentlemanly manners and practical knowledge peculiarly fit him for this situation. Neatness and order prevail in every department of his boat." Reed was an experienced captain, having previously skippered steamboats in the tricky waters of the Bay of Fundy. For his new assignment, Reed had only a compass and crude navigational charts available, although he had a pilot on board at all times. Most of the time, Reed and his pilot had to depend on visual sightings of the few lighthouses scattered along the coast and their knowledge of coastal ledges and currents to pilot the ship.

Drawing of *Royal Tar* by Sidney Winslow. *Courtesy of Vinalhaven Historical Society.*

The Circus and the Fire

In the nineteenth century, the traveling circus was one of the chief entertainments for rural North America. Throughout the summer, caravans of animals and performers crisscrossed the country, stopping for a performance or two before moving on to the next town fifteen or twenty miles away. Stuart Thayer writes in *Travelling Showmen*, "A circus in the 1830s might have six to ten wagons and a dozen ring horses."

The Macomber-Welch Circus Company was modest even by the standards of the day. In October 1836, the company was completing a seven-hundred-mile, three-month tour of New Brunswick and Nova Scotia. Included in the circus were a fifteen-year-old elephant named Mogul, a gnu, a zebra, two dromedaries and two pelicans. Wheeled vehicles carried a Bengal tiger, two lionesses, a leopard, a hyena and assorted smaller animals. The caravan also included a collection of birds and a variety of snakes. The troupe was exhausted after its long trip and was looking forward to returning to Boston. The fear of a plodding, overland trip spurred the circus's owners to rush to St. John to catch *Royal Tar*'s last run to Portland in late October.

To accommodate the circus, *Royal Tar* was modified in ways that modern safety regulations would never have permitted. For example, to make room for the animals, two of the four lifeboats were removed. A special boarding ramp was built for the elephant, which was placed on the deck in a stall directly over the ship's boilers. As we will see, this was to have serious consequences. Other animals were placed in cages below the deck.

Royal Tar's last voyage began on Friday, October 21, under fair skies, and the ship made its regular stop at Eastport that evening. Gale-force winds, however, caused the vessel to remain at Eastport over the weekend until the steamer was able to set sail on Monday morning, October 24. Captain Reed was only able to go another twenty miles until high winds forced him to anchor off Cutler Harbor. The next day, Reed, who was now three days behind schedule, weighed anchor and took the more sheltered route along the coast through the Deer Isle Thorofare and toward the Penobscot Bay Fox Islands of Vinalhaven and North Haven.

In attempting to reconstruct what happened next, it is important to bear in mind the words of Peter Dow Bachelder in his book *Shipwrecks and Maritime Disasters of the Maine Coast*:

> *Many of the facts concerning the* Royal Tar *during her final hours will never be known. Numerous details are scant or entirely lacking and those*

> *which exist, occasionally conflict with or contradict with one another. Because of this it is now impossible to fully reconstruct what actually happened…Despite their shortcomings the various eyewitnesses descriptions are invaluable because they leave no doubt about the utter confusion, terror and panic that gripped nearly everyone aboard* Royal Tar.

Early in the afternoon of October 25, *Royal Tar* proceeded across East Penobscot Bay toward Vinalhaven Island, buffeted by strong northwest winds. Without anyone realizing it, the boilers had run dry. The chief engineer had been up all night working on the engines and was asleep in his bunk. If he had told anyone to check the boilers, the message had apparently not gotten through.

When Captain Reed saw the steam pressure dropping, he anchored off Vinalhaven to investigate the problem. The red-hot boilers had already ignited the wooden wedges that had been placed under the elephant's cage for stability. The resulting fire spread quickly below decks. To make matters worse, the ship's firefighting equipment was located in the engine room near the boilers, where billowing smoke made them virtually inaccessible.

Captain Reed soon realized that there was little hope of putting out the fire and ordered the two remaining lifeboats put over the side. The larger of the two boats quickly filled and, blown by the wind, headed across East Penobscot Bay for Isle au Haut. With the ship incapacitated, Reed cut *Royal Tar*'s anchor line and raised the sails, hoping he could beach his ship. He then got into the smaller "jolly" boat, intending to ferry passengers to Vinalhaven, about a mile away. Strong offshore winds made this impossible.

In addition to Captain Reed's account of the fire, several passengers wrote of their experiences, giving us clues as to what may have happened. Herbert Fuller, manager of the circus, wrote a particularly grim description:

> *At this time a great many people jumped overboard and were drowned. The screams of women and children; the horrid yells of the men; the roaring of the storm and the awful confusion baffle description…I sat on the stern rail until my coat took fire. I fastened a rope to the tiller chain and dropped over the stern where I found about fifteen others hanging in different places, mostly in the water. While holding on I saw several drown.*

With the steamer now burning furiously, some of the animals had been pushed over the side and were swimming frantically around the stricken vessel. Most of the caged animals, however, were not let out. Mark Warner writes, "It was feared that in their panic, they would be more deadly to the passengers and crew than the fire and the sea. By that time, most of them had probably, mercifully, succumbed to the smoke."

Mogul the elephant was freed from his stall, although he resisted all efforts to get him over the side. A Canadian passenger, Stinson Patten, described the scene: "He remained, poor fellow, viewing the devastation, until, the fire scorching him, he sprang over the side and was seen striking out for shore with his trunk held high in the air."

Stories of Mogul leaping into the water and crushing some passengers on a makeshift raft are probably apocryphal. Roy Heisler at the Vinalhaven Historical Society suggests that when the elephant crashed through *Royal Tar*'s rail, he took several passengers with him. A newspaper report that Mogul had swum to a nearby island proved to be untrue. Several days later, his body was found floating south of Vinalhaven, near Brimstone Island, about six miles from where *Royal Tar* caught fire.

Rescue and Aftermath

As Captain Reed was loading survivors into the smaller lifeboat, he spotted the revenue cutter *Veto* in the distance. *Veto*'s captain, Howard Dyer, a Vinalhaven resident, had seen the fire and was coming to investigate. Dyer approached *Royal Tar* cautiously, however, since *Veto* was carrying a large quantity of gunpowder. Reed rowed his lifeboat toward the cutter with a number of survivors, beginning the first of several transfers that ultimately resulted in the saving of forty lives. (Circus manager Herbert Fuller was one of those rescued by Reed.)

Four hours after the fire broke out, William Marjoram, a passenger who had already been rescued, commandeered Captain Dyer's gig and rowed back to check for any remaining survivors. Marjoram wrote, "On reaching the wreck there was a woman holding on to the bowsprit with a child in her arms and another in the water with her clothes burnt off holding on to a piece of rope. She drowned before I could get to her." Marjoram was able to save the other woman, but not the child in her arms.

In all, thirty-two people died in the catastrophe: twenty-nine passengers and three members of the crew. Sixteen people in the large lifeboat made it safely to Isle au Haut late in the afternoon of October 25. Later that evening, *Veto* ferried the remaining passengers to Isle au Haut, where a survivor wrote, "There they were treated with much kindness." The next day, most

Revenue cutter *Veto* is seen approaching *Royal Tar*. Captain Reed, wearing a black hat, is picking up survivors while others cling to the burning ship. *Painting by Stephen Busch.*

of the passengers and crew boarded a schooner, which took them across the now calmer waters of Penobscot Bay to Rockland. From there, they were transferred to Eastern's steamer *Bangor*, which took them to Portland.

The remains of *Royal Tar* sank, and the hull has never been recovered. Mark Warner tells us that in 1962, divers discovered a melted winch and a piece of rudder. A 1992 clipping from the *Castine Patriot* shows a picture of several marine items, including a rudder recovered by two fishermen; these objects may have come from *Royal Tar*. The Vinalhaven Historical Society has a clipping that says that in 1935, William Combs succeeded in salvaging *Royal Tar*'s anchor and sold it for $400. And in 1938, the late Philip Sawyer found a bone, which appeared to be the lower part of an elephant's leg.

Royal Tar was not insured, and the company's loss was estimated at $50,000. All of the cash from the circus tour was lost, in addition to the value of the animals (the elephant alone was worth $15,000), the wagons and other gear. In addition to their luggage, several passengers lost considerable amounts of money, perhaps as much as $15,000. In all, the monetary loss of the disaster was placed at $200,000.

Captain Ezekiel Jones of the U.S. Revenue Service was sent to investigate the affair, and in the following weeks, he visited Vinalhaven, Isle au Haut and Deer Isle, where, in the words of Peter Bachelder, "he and his men searched the shorelines, interviewed several people and took depositions." Captain Dyer of *Veto* was at first criticized for his actions, but after talking with survivors, Captain Jones heard nothing but praise. "I have talked with many of them [*Royal Tar* survivors] and they all look upon him as their preserver." In fact, Dyer had brought *Veto* dangerously close to the blazing steamer, receiving burns himself, in an effort to save passengers. Howard Dyer continued as captain of *Veto* until 1843, when he became keeper of Brown's Head Light at the western entrance to the North Haven Thorofare.

Captain Thomas Reed also received high praise from the survivors, who "gratefully acknowledged his extraordinary exertions and perseverance, in saving the lives of his passengers while in such imminent danger at the time the boat was on fire." Upon his return to St. John, Captain Reed was hailed as a hero. Realizing that he had lost all of his possessions on the ship, his friends presented him with $750, "as a mark of respect for his exertions to save the lives of passengers and crew of *Royal Tar*." On March 12, 1838, Reed was given command of a new steamship, *Nova Scotia*, serving ports on the Bay of Fundy. Thomas Reed ended his days as harbormaster of St. John.

For many years after the tragedy, rumors circulated about lions and tigers seen roaming nearby islands. And it is said that Crotch Island, south of Deer Isle, was infested with exotic snakes. The December 1, 1944 *Rockland Courier-Gazette* ran a Sidney Winslow story, quoting citizens of "high repute" who would take an oath that they had witnessed the spectacle of a mysterious flare in the vicinity of the spot where the old circus ship met with her untimely end. As with much of what happened during *Royal Tar*'s final hours, these make good stories, though they are not supported by any factual evidence.

Now on Stormy nights in October the fishermen up Pequod way
Tell of a ghostly light out on the eastern bay.
Perhaps this shimmering radiance, as it lights the waves afar,
Is the soul of the ship returning,
The ghost of the "Royal Tar."

—*Anon*

THE WRECK OF *CASTINE*

Ninety-nine years after the burning of *Royal Tar* off the eastern side of Vinalhaven, the sixty-nine-ton steamer *Castine* ran on a ledge off the western shore of the island. The accident, which occurred on June 8, 1935, was called by the *Rockland Courier-Gazette* "one of the worst steamboat disasters recorded in the history of Penobscot Bay." Four people perished in the tragedy.

Castine was an old ship, built in 1889, though her co-owners, Captain Leighton Coombs and his brother and chief engineer Perry Coombs, kept her in immaculate condition. She was seventy-one feet long with a fourteen-and-a-half-foot beam. In 1935, *Castine* was based in Belfast and used on the Camden-Islesboro-Belfast run, as well as a pleasure boat for special events. Not surprisingly, *Castine* had no radar, which was in the early stages of development.

Castine left Rockland that Saturday in June carrying seventy-five passengers, most of whom were members of the Limerock Valley Pomona Grange heading for their annual outing with farmers on Vinalhaven and North Haven at the Pleasant River Grange. An hour and a half later, *Castine* was about two miles outside Carver's Harbor when she struck a ledge in the thick fog.

Castine was built in Brewer in 1889 and was seventy-one feet long. The ship was said to have been kept in immaculate condition. *Courtesy of Vinalhaven Historical Society.*

Castine *Didn't Sink: "It Just Tilted"*

Accounts of the accident vary, although everyone agrees that because of the fog, Captain Leighton Coombs had slowed his vessel before it hit the ledge. According to a newspaper account, *Castine* ran onto Bay Ledge off Vinalhaven with sufficient impact "to send some people instantly into the cold waters." Those inside were thrown from where they were sitting around the cabin. The collision tore a large hole in the starboard side, filling the cabin and boiler room with water.

Alvin Rackliff, age fifteen at the time, had a slightly different version of the accident. Alvin was sitting up in the bow and related that they had just passed the steamer *North Haven*, going the other way, before hitting the ledge:

> *I could see an apron of kelp on top of the water, and I thought there must be something wrong. Then the boat slid up on top of the ledge and didn't tilt much at all. In a minute somebody came up from the engine room and said, "everybody to the high side so we can right her up." When he said that all the ladies and older folks in the back of the boat went to one side and of course she started leaning. Everyone pressed against the railing and pretty soon it broke and all those people went overboard.*

F.L. Morse, superintendent of schools in Thomaston, was one of those dumped in the water. Morse was sitting on a deck chair when *Castine* tipped sideways, throwing him and his chair overboard. "There he was in his blue serge suit, flopping around," young Rackliff said. "He couldn't swim, but he managed to stay afloat until they could get him out."

Based on a conversation he had with Captain Coombs, John Moore of Rockland provides us with a third view. Moore, who was standing next to the pilothouse, remembers asking Captain Coombs, shortly before the accident, how far they had to go. The captain's response was, "We are about halfway." Coombs then told Moore, "If we don't hear that bell pretty soon I don't know where we will be." He asked me to let him know if I saw any lobster pots on our port side, which would indicate shallow water, Moore said. "I told him we were among lots of pots. At that point he signaled half speed but did not stop the boat. Almost at once we piled on the ledge."

Moore later commented that the captain should have stopped the vessel when he realized that they were surrounded by lobster pots. He felt that had Coombs stopped *Castine*, they might have drifted on to the ledge but without causing any great damage.

In John Moore's opinion, *Castine* was off course when they hit the rock. Moore was standing beside the port window of the captain's cabin at the time of the accident. "Coombs had the window partly open, and he seemed to be looking at a note book from time to time, but I didn't see any charts." Incidentally, Moore's wife was one of those thrown into the sea when the ship ran aground. She grabbed the rail, which gave way, and was thrown into the sea. Her husband immediately jumped in and pulled her out.

Another passenger was Ruth Lucas, then Ruth Freeman, age nine. Lucas, now in her eighties, has vivid memories of the accident. Ruth was with her baby brother, Richard; her mother, Gertrude Freeman; and her aunt Laura Ingraham. All were looking forward to visiting Gertrude's father, "Grampy," who lived on Vinalhaven. As Ruth's father drove them to Rockland from their home in Camden, her mother and aunt looked out at the fog and were reluctant to go. Ruth persisted. "You promised me mother," she remembers saying.

The Freeman family, who were not members of the grange outing, were in the cabin when *Castine* struck the ledge. Ruth's mother and aunt immediately went on deck. When *Castine* lurched again, her mother and seven-month-old brother Richard were thrown into the water. Crew members threw life jackets to people in the water, and then—"seemingly from out of nowhere," Ruth said—a rope was lowered to her mother. "My mother couldn't swim a stroke, but she hung on to Richard with one arm and the rope with the other arm. The lord was looking after us," Ruth told me.

Ruth confirmed the fact that *Castine* didn't sink: "It just tilted." The resulting list, however, threw a dozen passengers into the water. (The captain later said it was the panicky rush of the passengers to starboard that also opened the seams of the elderly vessel, causing her to partially submerge.) "When water started pouring in there was lots of screaming and yelling," Ruth said. People were in shock, but she kept wondering why someone didn't jump into the water to try to help her mother and baby brother. Help finally arrived when Vinalhaven fisherman Hanley Dyer rescued her mother and brother, as well as others still in the water.

Ruth herself had never learned to swim. She remained on board until a lifeboat from the North Haven ferry rescued her. Everyone was given lifejackets, but *Castine*'s only lifeboat leaked so badly when it was launched that it sank. Ruth recently told me that her mother had almost drowned when she was a girl so she never let her children go in water above their knees. "I guess she was overly protective."

The Rescue

When Captain Coombs realized his ship was on a ledge, he sounded a distress signal of four long blasts on the horn repeated continuously. The emergency call was heard by Vinalhaven and North Haven fishermen in the area, as well as on the steamer *North Haven*, headed for Rockland. Help was on the way, but *Castine* was not easy to find. In the dense fog, it took rescuers over an hour to locate the stricken vessel.

When the steamer *North Haven* finally arrived on the scene, Captain Roscoe Kent sent over lifeboats and began to take passengers off *Castine*. (Apparently, the fog was so thick that Captain Kent could barely see the stricken vessel.) At about the same time, Vinalhaven fisherman Hanley Dyer arrived and, as noted, rescued Ruth's mother and baby brother, who had been in the water for over an hour. Nine-year-old Ruth remembers thinking that her brother and mother survived because they were near the partially submerged smokestack, which warmed the water.

Colonel Fernando Philbrick was a ninety-year-old veteran of the Civil War. A native of Hope, Maine, Philbrick enlisted in the Union army at the age of seventeen and saw action at Port Hudson, Louisiana. "Six hundred men were killed during the attack, but I only got a scratch." Colonel Philbrick said that when *Castine* lurched on her side, he was thrown to the deck and pinned under a chair. Although the water was rising, Philbrick insisted that the badly injured Mrs. Grace Packard be taken off the ship before he was attended to. "When I first saw her, I thought she was dead." Witnesses say

Some *Castine* survivors about to be picked up. *Courtesy of Vinalhaven Historical Society*

he shouted, "Get the woman first, I'm all right!" Philbrick was taken to Vinalhaven and later transported across the bay to his home in Camden. "It was just 'another tight place' for the keen-witted veteran who has known many of them in his life," a newspaper account noted.

Merle Mills was in his boat on the western shore of Vinalhaven when he heard the distress signal. After locating *Castine*, Mills filled a lifeboat with passengers and towed them to the ferry *North Haven* waiting nearby. He then returned to *Castine* and removed the remainder of the crew, Captain Coombs being the last to leave the ship. Mills then transferred them to a Coast Guard patrol boat that had just arrived from Rockland.

Although there were many acts of heroism, Captain Leighton Coombs singled out the work of William Roberts of Belfast, a member of his five-man crew. "Roberts saved eight persons from the water himself. His work was remarkable," Coombs said. He also spoke warmly of the help given by Captain Roscoe Kent and the crew of *North Haven*, as well as the aid of Hanley Dyer and Merle Mills.

Unfortunately, not all the passengers survived. Mr. and Mrs. Charles Wooster both died in the disaster. Mrs. Wooster, sixty, was thrown overboard and drowned before she could be rescued. Mr. Wooster, sixty-three, hung on to a stanchion rail for a while before he collapsed and died of a heart attack. Two other people, Evelyn Bartlett of Washington and Rebecca Alley of Camden, died in the Knox Hospital in Rockland within a few weeks of injuries received in the wreck. Both were members of the Limerock Valley Pomona Grange.

"We Loved Castine *as One of the Family"*

Most of the survivors were put on board the steamer *North Haven* and taken to the Rockland ferry dock, which today is the Rockland Fish Pier. There they were greeted by a large crowd of friends and relatives who had learned of the disaster through telephone calls from Vinalhaven. Ruth Lucas says she doesn't remember landing in Rockland, probably because she was still in a state of shock.

When her father, who was on the mainland, heard about the accident, he was so worried that he hitched a ride on the Coast Guard patrol boat heading out to the scene. In the meantime, his wife and young son had been taken to Vinalhaven to warm up. They returned to Rockland later in the day on the Vinalhaven ferry. Ruth said she remembers that night when the doctor came to the house to check on her mother and gave her some brandy. "I remember that because my mother was not a drinking woman," she said.

Although the heavy fog remained, it did not deter swarms of fishermen and others from nearby islands from flocking to the stricken vessel and taking everything removable. The *Courier Gazette* reported, "Men brought pinch bars, saws, axes and other tools to quickly take apart all that could be removed. Small boats were soon laden with pieces of decking and sides of the small steamer." A day later, the newspaper reported that nothing was left except the stripped hull.

Were the looters also looking for women's purses? If so, they were disappointed. Before she was taken off *Castine*, Alvin Rackliff's mother, Lillian, gathered up all the pocketbooks she could find and put them in a bag, telling her son to "guard it with his life." Ruth Lucas verifies this report, saying that "women's pocketbooks were not stolen; they were retrieved." Indeed, a week later, Merle Mills found a pocketbook floating in the water off North Haven belonging to Emma Kinney. Mills dried out the contents and mailed them to Kinney in St. George.

Propped up in bed, his legs swathed in bandages, Perry Coombs, *Castine*'s engineer and brother of Captain Leighton Coombs, spoke to the press the next day from his home in Belfast. He reported that his brother was too weak from exposure and the mental strain to talk, but "he would rather have gone

Castine's hulk was prey to looters and salvagers who began to tear the remains of the vessel apart. *Courtesy of Vinalhaven Historical Society.*

Castine's hull on Cedar Island today has long been used as a guesthouse by the Baker family. *Author's collection.*

down with her in thirty fathoms than have this happen. We all loved *Castine* as one of the family." With nothing left of their ship but the stripped hull, the brothers did not attempt salvage.

Today, passengers on the Vinalhaven ferry pass close to Cedar Island, where the forward part of *Castine*'s hull, bottom side up, is clearly visible. For many years, the Baker family, who own Cedar Island, used the hull as a guesthouse. It sits there, still intact, less than a mile from where the wreck occurred.

Hit By Lightning: A True Story

The weather report that Friday night in August called for "thunder showers, ending by morning." It also mentioned "possible lightning strikes." Now, I've always put lightning strikes in the category of shark attacks. You hear about them, but they rarely happen, especially to you.

Suffice it to say, the lightning storm we experienced on Vinalhaven that night was the worst I had ever seen. As I looked out our bedroom window, the night sky was continuously illuminated for several hours. With lightning strikes

Sybaris before the storm. *Author's collection.*

hitting seemingly everywhere, it was a scene reminiscent of what a World War I battlefield must have looked like. A neighbor's chimney was hit, as were several boats in Carver's Harbor, and that was just in our immediate area.

The previous spring, I had become the proud owner of a twenty-six-foot Sisu named *Sybaris*, which I purchased from a dealer in Rockport. I was becoming familiar with the multipurpose boat, which had been designed thirty years previously to serve everyone from offshore fishermen, to commercial lobstermen, to pleasure boaters like myself. Sisu is a Finnish word that, loosely translated into English, means to "to have guts." As we shall see, this was an appropriate designation for the trials my boat was about to face.

The next morning, I rowed out to inspect *Sybaris*. From a distance, I thought everything looked OK. As I climbed aboard, however, I noticed plastic shards scattered around the cockpit, and my heart began to sink. Further inspection revealed that they were the remains of the VHF radio. When I looked up and saw what was left of the radio antenna split in several directions, I realized that my boat had indeed been hit by lightning. This was confirmed when the rest of the electronics (GPS, radar, Fishfinder, Smart Pilot) did not function; I was also unable to start the engine. Although I didn't realize it at the time, I had taken my last ride on *Sybaris*.

Confronted with this disaster, I tried to figure out what to do next. Since it was Saturday morning, I couldn't inform the insurance company until Monday. Therefore, I contacted our caretaker, Pete Gasperini, hoping that he might be able to get it started. After confirming the fact that the electronics were shot, Pete informed me that the engine had also been ruined. At that point, we weren't sure what had caused it to fail, though it certainly appeared to be connected with the lightning strike. The good news was that although the engine wouldn't start, at least there was no exit hole in the hull, which would have caused the boat to sink. I should add that the day before, I had taken some friends to Rockland in *Sybaris*, with no indication that the engine (a Crusader Marine V-8) wasn't in perfect condition.

I checked my insurance policy and found to my relief that "electronic navigational equipment was not subject to a deductible." Nor was there anything specific regarding the engine in a paragraph titled "Exclusions." "Don't worry," Pete reassured me, "the strike probably knocked out the starter, which is easy to replace."

My next step was to contact Thayer's Y-Knot Boatyard on North Haven and explain what had happened. Co-owner Collette Haskell, who was very

Lightning strike. The storm that night looked like a World War I battlefield. *Wikipedia.*

accommodating throughout the whole process, told me to have *Sybaris* towed over, and their mechanic would check it out. At that point, I remember optimistically thinking that the boat would probably be laid up for a week or two while we replaced the electronics and fixed the starter.

In fact, it was a good deal more complicated than that, but I am getting ahead of myself. Monday morning, I called the insurance company. The agent in charge of the local office on Vinalhaven told me that her son had been on the phone Friday night when the line was hit by lightning, and he had received a terrific shock. He was all right, but it helped put what happened to my boat in perspective.

By mid-week, I had received a letter from the insurance company with a list of questions to answer. A representative from the company called, and after hearing my explanation of what had happened, he assured me that my losses would be covered but implored me "to be patient." The remainder of the summer passed as Thayer's Boatyard was understandably reluctant to repair my engine without authorization from the insurance company. I hesitated, too, thinking of the expense involved. After weeks of negotiations, however, we finally received permission to proceed. The findings were not encouraging.

The key question was whether the Crusader engine had been damaged goods *prior* to the lightning strike. When Thayer's mechanic finally received permission to pull the engine, he found "extensive corrosion and heavy rust encasing the crank," leading him to believe that the engine compartment had been filled with water at some time previously. In turn, this raised the question: was the damage to the engine a result of the lightning strike, or was it just a coincidence that the engine seized up when it did?

Thayer's report to the insurance company was appropriately ambivalent: "It is undeterminable whether the damage to the engine is connected to the lightning strike or is just an untimely, unfortunate coincidence." After a twenty-minute examination of the boat, however, a marine surveyor hired by the insurance company informed them that the damage to the engine was *not* due to the lightning strike.

Naturally, I disagreed with the surveyor's conclusions. Prior to purchasing *Sybaris*, it had been inspected by a licensed surveyor. Furthermore, the Crusader engine had performed without incident since I had bought it, including the trip to Rockland the day before the strike. I asked myself what could have caused the engine to suddenly seize up except the lightning strike? Apparently, salt water had been sucked into the engine through the manifold during the strike and had remained for several weeks. Wouldn't this

A family outing in *Sybaris*. This photo was taken the day before the lightning strike. *Author's collection.*

explain the rust and corrosion buildup? In spite of my protests, the insurance company agreed with the findings of its marine surveyor. In mid-October, I was sent a check for the electronics but not a penny to replace the engine, which was a total loss.

Conclusions? Read the small print in your policy before getting your hopes up. Mine stated: "We will not provide coverage for any loss or damage caused by, or resulting from wear and tear, lack of maintenance, corrosion or deterioration." Based on the information provided by the marine surveyor, the company decided that the damage to the engine was caused by *wear and tear/corrosion*, not by lightning.

Finally, when the weather report calls for "possible lightning strikes," take it seriously and remove, or at least lower, the radio antenna, which on my boat was the principal conductor. It might also be worth moving your boat to a less exposed anchorage. Lightning strikes can indeed occur, even to you!

Bibliography

Bachelder, Peter Dow. *Shipwrecks and Maritime Disasters of the Maine Coast.* Portland, ME: The Provincial Press, 1997.

Bennett, Frank M. *The Steam Navy of the United States*. Pittsburgh, PA: Warren & Company, Publishers, 1896.

Black, Frederick Frasier. *Searsport Sea Captains*. Searsport, ME: Penobscot Marine Museum, 1960.

Cotterly, Wayne. *Hurricanes and Tropical Storms: Their Impact on Maine and Androscoggin County*. On-line edition, 1996.

Davis, Charles G. *Around Cape Horn*. Edited and introduced by Captain Neal Parker. Camden, ME: Down East Books, 2004.

Duncan, Roger F. *Coastal Maine: A Maritime History*. Woodstock, VT: Countryman Press, 2002.

Fitts, James H. *Lane Genealogies*. Vol. 3. Exeter, NY: The News-Letter Press, 1902.

Gross, Clayton H. *Island Chronicles: Accounts of Days Past in Deer Isle and Stonington*. Stonington, ME: Penobscot Bay Press, 1977.

Ludlum, David M. *Early American Hurricanes, 1492–1870*. Boston: American Meteorological Society, 1963.

Paine, Lincoln P. *Down East: A Maritime History of Maine.* Gardiner, ME: Tilbury House, 2000.

Rowe, William Hutchinson. *The Maritime History of Maine*. Gardiner, ME: The Harpswell Press, 1989.

Shain, Charles, and Samuella Shain, eds. *The Maine Reader: The Down East Experience from 1614 to the Present.* Boston: David Godine, Publisher, Inc., 1991.

Sharp, Jim. *With Reckless Abandon.* Rockport, ME: Down East Books, 2007.

Snow, Edward Rowe. *Storms and Shipwrecks of New England.* Beverly, MA: Commonwealth Editions, 1973.

Snow, Ralph Linwood, and Captain Douglas K. Lee. *A Shipyard in Maine.* Gardiner, ME: Tilbury House, 1999.

Thayer, Stuart. *Travelling Showmen: The American Circus Before the Civil War.* Detroit, MI: Astley and Rickets, 1997.

Thorndike, Virginia. *The Arctic Schooner Bowdoin.* Unity, ME: North Country Press, 1995.

Van Horn, David. *A Brief History of Penobscot Bay.* Castine, ME: Robert's Press, 2003.

Warner, Mark. *The Tragedy of the* Royal Tar. Newcastle, ME: Warner Publishing, 2010.

Williams, Tony. *Hurricane of Independence.* Naperville, IL: Sourcebooks, Inc., 2008.

Archival Materials Consulted at the Following Locations

Bowdoin College Library Special Collections, Brunswick, ME
Deer Isle–Stonington Historical Society, Deer Isle, ME
John Fitch Steamboat Museum, Warminster, PA
Maine Historical Society, Portland, ME
Maine Maritime Academy, Castine, ME
Maine Maritime Museum, Bath, ME
Mariners' Museum, Newport News, VA
Penobscot Marine Museum, Searsport, ME
Portland Public Library, Portland, ME
Vinalhaven Historical Society, Vinalhaven, ME
Vinalhaven Public Library, Vinalhaven, ME

About the Author

Courtesy of Christopher Krueger.

Harry Gratwick is a lifelong seasonal resident of Vinalhaven Island in Penobscot Bay. A retired history teacher, Gratwick had a forty-six-year career as a secondary school teacher, coach and administrator. He spent most of these years at Germantown Friends School in Philadelphia, Pennsylvania, where he chaired the History Department and coached the baseball team.

Harry is an active member of the Vinalhaven Historical Society and has written extensively on maritime history for two Island Institute publications, the *Working Waterfront* and *Island Journal*. *Stories from the Maine Coast* is his fourth book about Maine.

Gratwick is a graduate of Williams College and has a master's degree from Columbia University. Harry and his wife, Tita, spend the winter months in Philadelphia. They have two grown sons, a Russian daughter-in-law and two grandsons.

Visit him at www.harrygratwick.com.

Also by Harry Gratwick:

Penobscot Bay: People, Ports and Pastimes
Hidden History of Maine
Mainers in the Civil War